Hattie

She Was Wired Differently

LOIS REIMERS

Outskirts Press, Inc.
Denver, Colorado

The opinions expressed in this manuscript are solely the opinions of the author and do not represent the opinions or thoughts of the publisher. The author has represented and warranted full ownership and/or legal right to publish all the materials in this book.

Hattie
She Was Wired Differently
All Rights Reserved.
Copyright © 2011 Lois Reimers
v2.0

This book may not be reproduced, transmitted, or stored in whole or in part by any means, including graphic, electronic, or mechanical without the express written consent of the publisher except in the case of brief quotations embodied in critical articles and reviews.

Outskirts Press, Inc.
http://www.outskirtspress.com

ISBN: 978-1-4327-7405-9

Outskirts Press and the "OP" logo are trademarks belonging to Outskirts Press, Inc.

PRINTED IN THE UNITED STATES OF AMERICA

*Dedicated to my son Thomas Reimers
who's encouragement and support for my labor of love
accomplished its mission.*

TELL ME A STORY

By Lois Reimers

Long ago there were storytellers in villages
towns and the wilderness's everywhere
sitting under shade trees and in front of
fireplaces, telling and retelling
      the history of their people

The little children would be the most fascinated
      listeners,
wanting to hear the stories of long ago,
      told over and over again.

They loved best the stories of their
Mothers, Fathers, and Grandparents.

Those storytellers of long ago have almost all gone now.
Let us give our children the feeling of belonging
      once again and the warmth which come from
          hearing the stories of their families' struggles

Both the failures and successes.

Introduction

I know my Great Aunt Hattie Fuller would be appalled that I have written her life story, but I think it is time for it to be told. The point is not to "out" her for some sick, unkind reason, but rather because I felt if someone who lived in her era (1876 to 1946) wanted so badly to live as a different gender, then it could not have been a preference.

I ran across her story while tracing my genealogy. My grandmother, Nellie Fuller Carey, had me type her family's early history. Her family lived on an Indian reservation while she was growing up in South Dakota. She also said her father, Melroy Fuller, could be traced back to the Mayflower Fullers, Samuel and Edward.

I decided I would try to make this connection to the Pilgrims after my grandmother passed. I then ran into this fascinating story of my grandmother's younger sister, Hattie. I learned she had lived almost her entire adult life as a man. She dressed as a man, smoked cigars, and was married to a woman by the name of Inez. They lived together in the small town of Rhame, North Dakota. Hattie had bought the weekly newspaper and operated it from 1916 to 1946. She was on the school board and belonged to several men's clubs in town, where she was known as Cappie or Alfred D. Fuller.

I decided to tell the story of this remarkable woman and her family, not to out her, but rather to show how deeply the choices of who we are must be etched on our psyche.

As you read this book, you will wonder how Hattie found the courage to lead the kind of life she led. Thank God she had the family, especially her father, Melroy Fuller, who supported her all her life. He shielded her secret by accepting her as exactly the way she was.

Lastly, you may ask why I am outing her now.

I thought more people need to look at the struggle that still exists for a person who is perhaps wired differently.

I asked, Can we begin to look with love and tolerance at people who are different? I asked myself, Isn't that what all religions say to do?

Then I asked why should people need to hide who they are?

I asked myself what God would have me do.

Acknowledgements

First, I need to acknowledge my grandmother, Nellie Louise Fuller, Hattie's oldest sister. In 1945, she asked me to type her story about growing up on an Indian reservation and an Indian agency around Chamberlain, South Dakota. The time period spanned from 1880 to 1890—ten long and interesting years, she said.

Typing her story seemed to give me the imagination to continue seeking stories of ancestors. It piqued my interest in the study of genealogy. This is how I stumbled onto Hattie's story.

Next, I need to acknowledge my wonderful, hard-working, open-minded parents.

They loved me and supported me in everything I undertook—from joining the Women's Army Corps during World War II at age nineteen, to helping my husband raise our four sons. My mom and dad, Ethyl and Jerry Tolliver, encouraged and loved me. They taught me how to work and love. I bless them every day of my life.

Then there is my wonderful, supportive husband, Jack. He has never uttered a word against the time I take to follow the passions of my heart, which are many: from writing to painting to my dedication to all my friends and family. Never a cross-word, ever! Bless him; a greater gift I could not have received from God.

Also, I must thank Merrie Sodder, Fred Fuller's granddaughter, who found me online many years ago. Without her help in digging up and sending me all the documents of our mutual ancestor I would have been unable to tell her story.

Lea Reimers, my daughter-in-law, needs to be mentioned and thanked for her very valuable contribution to the book. She holds a master's in English and I turned this manuscript over to her for corrections.

Then I need to give thanks to my very helpful Thursday writers group. Its members encouraged me and corrected my spelling and grammar mistakes with such kindness. Special thanks to Laurie Greene, Joe Minton, and Mirian Bethancourt.

To my two dear friends Al and Ruth Bendekgey. Al for his expertise with a camera, and Ruth for the hours spent away from her wedding business to help me. Thanks friends!

And finally to you, dear reader, who had the courage and curiosity to want to understand transgender people. They are people who live among us and have been so discriminated against that they have a high rate of depression, suicide, substance abuse, and relationship difficulties. I applaud your open-mindedness.

As I researched my Aunt Hattie's life, I applauded her for her pluck, which her mother, Elizabeth, called "guts." I acknowledge most of Hattie's family and especially her father, Melroy, my great-grandfather, for his unwavering love of Hattie—accepting her exactly as she was and protecting her with his love. That I acknowledge most!

About the Author

I have been a writer of stories since I was a little girl. It seemed to come naturally for me to tell stories on paper. Each time I pick up my pen, I seem to become Hattie.

In 1945, I became interested in genealogy while typing my grandmother's life story. She had a grand story to tell of growing up among the Sioux Indians on an Indian reservation in South Dakota.

My grandmother had a younger sister named Hattie. I stumbled onto the family secret that Hattie had lived almost her whole adult life as a man. I became fascinated when I inherited all the family pictures, including one of Hattie and the women she married. I was captivated; my research began, and then the story had to be told.

Passion for this story has kept me writing for seven years. I have joined a few writing groups. In 1980, I authored a book called *Our Child's Root's*. I have had a short story and a poem published. I feel after many rewritings, it is time to publish this timely story.

Contents

INTRODUCTION...vii
ACKNOWLEDGEMENTS ix
ABOUT THE AUTHOR..xi

CHAPTER ONE.. 1
"Where Am I?"
First Memory – Fred Born

CHAPTER TWO 9
Going West – Chamberlain, S.D.
War Dance

    Tintype Fuller Family 1881 22
    Picnic Fuller Family 1887...22
    Indian Language Newspaper 23
    "Grim" Fuller Family 1890......................................24
    Salem, South Dakota 1891.......................................24
    Melroy Fuller Family Page.......................................25

CHAPTER THREE 27
Prairie Fire
Half Moon

CHAPTER FOUR 45
Killing of Steers
Baby Ethel Arrives

CHAPTER FIVE 59
Rose Bud Uprising

CHAPTER SIX 69
Salem, South Dakota

CHAPTER SEVEN 83
Mitchel, South Dakota

    Hattie Age 26 92
    The Wedding 93

CHAPTER EIGHT 95
Chicago 1902
Hattie meets Inez

CHAPTER NINE 105
Mama's Death

CHAPTER TEN 117
Cappie meets Mel

CHAPTER ELEVEN 127
Wedding

CHAPTER TWELVE 139
Rhame, North Dakota
WWI

CHAPTER THIRTEEN 147
Fred's Visit

CHAPTER FOURTEEN 155
Health Problems

CHAPTER FIFTEEN 163
Papa Dies

CHAPTER SIXTEEN 173
The End, 1946

    Census 1920 183
    Papa's Printing Office 183
    Papa's Note..184
    Cappie's Visit to Salem, S.D. 185
    Census 1930 185
    Papa with Author "Lois" 186
    Melroy Fuller Death Certificate...................................187
    Melroy Fuller Newspaper Obituary 188
    Alfred D. Fuller "Hattie" Death Certificate 189
    Hattie Fuller "Alfred D. Fuller" Newspaper Obituary...............190

CHAPTER ONE

Where am I? What's going on? I can't talk! I feel so strange! Not exactly pain, more a sense of pain somewhere nearby. I feel detached from my body! What is happening? "Calm down, calm down!" I tell myself. I can hear my heart racing. Whatever's happened, panic surely won't help! I must get myself under control. My eyes and ears seem to be working. It looks and sounds as though I am on a moving train. How did I get here and why am I lying down? Wait, I hear Carl Peterson's voice. He's saying something about a stroke and the hospital in Baker, Montana. Is that what happened? Did I suffer a stroke? I can't seem to remember.

That must be it. Doc Weber had warned me last time I saw him that my time was nearly up, and a stroke was as likely as a heart attack. I'd asked him if there was any difference between the two deaths, thinking they were pretty much the same thing. He had stroked his short beard with one hand, a nervous habit of his, and said, "Well, if you have a heart attack, Cappie, given your long-term heart problems, it will likely be over very quickly. A stroke, on the other hand, would leave you partly or completely paralyzed for a time; at first, some people lose their ability to speak, see, or even hear."

He also had warned me to "put my affairs in order" and said, "perhaps, Cappie, you should take a room here in Baker." Guess that was

HATTIE

his way of saying it might be safer for my "secret" if I were to be closer to Baker and him. Doctor Weber is not one to mince words, but a good man just the same tells the truth painful or not. I am sure he will keep his promise to guard my secret. I know if any other doctor were to be there in my last mortal moments, it would bring on a terrible scandal. That had been my one fear all these years, a scandal that could devastate me and my family who had openly associated with me and guarded my secret. I had worked hard for many years not to disgrace my family because of the choices I had made.

Hearing Carl's voice was a relief. He was my good friend and never asked me why I was so adamant about no medical treatment anywhere but in Baker, or why I wanted him to accompany me personally. He never was one to judge or question; he just gave me one of his lopsided, toothy grins and said, "Ya, I will take you." I knew I could count on Carl. We had been friends for over twenty-five years. In the little town of Rhame, North Dakota, we helped each other out through all our personal sorrows and illnesses. I couldn't have asked for a better friend; if only I could tell him my secret now. How shocked even he would be to find his longtime friend A. D. Fuller was actually Hattie Fuller, a woman!

I've acted my role for so long; it is even hard for me to remember when I wasn't Alfred D. Fuller, or "Cappie" as my friends call me. Even my family, my late father, my sisters, and my only brother had become accustomed to calling me "Cappie." I smile to myself remembering their children calling me "Uncle Cappie." Of course, the children, too, like Carl, did not know my secret. Hattie was just a dim memory even to me now.

My sisters, Nellie, Marble, Elizabeth, and Ethel, their husbands, and my only brother Fred, and his wife, knew of me masquerading as a man, but none of their children knew. They called me "Uncle Cappie." I had been a family secret for so long now, let's see. I pondered as my woozy head allowed me to think in waves. I began my disguise as Alfred in 1903, and here it is 1946: *there, I remembered the date!* I must be better, I thought.

SHE WAS WIRED DIFFERENTLY

Who am I, I ask myself, and how did I get to be who I am now? Did some event change who I was while I was growing up? I had asked myself this many times. I wondered whether it was the semi-nomadic life Father led us through. Where and when did Cappie begin and Hattie end? Do I have time enough left to figure it out? All these thoughts seem to be racing through my head as I feel the motion under me of the train on its tracks. Clicky clack, clicky clack.

I can't seem to remember today or yesterday, I thought, but the past is vivid in my memory. I remember back before I turned four, when the world seemed such a wonderful, fascinating place, a place where Mama and Papa kept us safe and warm. I felt as though nothing could hurt us as long as they were there. By the time I was born, Mama and Papa had moved twice. They were living in Spring Valley, Minnesota, in 1872 when Nellie was born, then moved to Sleepyeye, Minnesota, in 1874, where Mabel was born. They were living in Sheffield, Iowa, at the time of my birth in 1876. Fred was born in Mason City, Iowa, although purely by chance, Mama said. Our family had gone for a Christmas visit to Mama's brother and his family, and we were stranded by a huge blizzard. By the time the trains could move again it was February and Fred had made his arrival.

I was only four when Fred was born in 1880, but young as I was, I could tell my Papa was very happy at finally having a son after three daughters. I so adored our Papa that I was happy for him. Everyone seemed to adore Papa. He was tall, handsome, and full of energy. He seemed to know everything and everybody.

I knew Papa always gave his first attention to Mama when he came home, then a hug and kiss for Nellie, as she was the oldest. I heard Mama tell everyone that Nellie was the apple of Papa's eye. I really didn't know what that meant, but I felt that it was a good thing, as Mama always said it very proudly, and she would put her arm around Nellie as she said it.

Then Papa gave Mabel a hug and kiss, saying, "How is my pretty lass today?" We all knew that Mabel was very pretty. More than once I heard people say as we were leaving church, "Oh, Mrs. Fuller, that

HATTIE

Mabel of yours is a beauty!"

Finally, Papa's attention would come around to me, the littlest and youngest until my brother Fred was born. I wanted to melt into the floor because I wanted his attention, yet I didn't want it either. I tried to hide behind a sister or a chair, thinking perhaps he would miss me, but he never did, "Come here, Hattie, my little intellect!" I didn't know what that meant either, but he said it proudly as he called me to him.

I remember moving over to stand between his bony knees and looking up into his blue, kindly eyes. "Well girl," he usually said, "What did you do today?" I tried to remember something that he would be proud of; I remember when I told him I had learned how to spell "God." He had looked very pleased and had patted me on the head and said, "All right, my dear, spell it for your Papa." I said, "G-O-D" in my quietest voice. He smiled that charming smile of his and said, "Good girl." Looking at Mama, he said, "She takes after you, Liz, very smart."

Their eyes met, and anyone could tell they loved each other very much; they had been married nine years when Fred was born.

Then I remember seeing his eyes seeking out baby Fred, lying in the cradle that he had built. Mama would rock it with her foot while sewing something with her always-busy hands.

Papa, his blue eyes twinkling, and with his duty to Mama, Nellie, Mabel, and me finished, went over and picked up baby Fred with his big, rough hands. His face softened, and his smile seemingly lit up the whole room. I could feel his happiness in holding his only son. I so wanted it to be me that he was holding and looking at with that kind of proud love. I remember the feeling of wishing to be the son that everyone seemed to love the best. Even my two sisters, whom the baby had also displaced as number one in Papa's heart, even *they* did not seem to care. But of course *they* had each other to play with, I thought, and I felt so alone. Did this have anything to do with my desire to be a male, I wondered?

We lived among the Sioux Indians while I was growing up and I

SHE WAS WIRED DIFFERENTLY

watched the Indian women constantly working; it seemed they were never done. The Sioux warrior way of life was hunting and fighting to protect and feed the tribe.

Lulled by the train motion, my mind began remembering roaming the prairies with Fred on our Indian ponies.

Fred and I felt part Indian ourselves and began thinking like them at times and taking on some of their ways. I must have been about ten and Fred was about seven when we began our roaming on the Lower Brule Indian Reservation across the Missouri River from the Black Hills Rose Bud Reservation.

The Sioux Indians on the reservation came onto the agency once a week to get their rations from the government. If they had no money, they traded all kinds of skins from different animals they had killed: buffalo hides, elk, skunk, coyotes, foxes and deer skins. Fred and I would enjoy looking over their kills. We watched them as they would demonstrate how they had killed the animals with their bows and arrows. They were proud of their hunting skills, letting out loud war whoops as they danced.

I felt the train rumble under me as I heard the train whistle screeching. This woke me out of my memories of those long-ago events. The shrill cry jolted me up out of the comfort of the passenger seat I had been lying on. The seat was tilted back so far that I was almost lying down. I looked down at myself and noted I had on my best waistcoat; my gold watch was snugly in its little pocket attached to the gold chain Inez had given me on our tenth wedding anniversary, many years ago. I was wearing my stripped blue-and-white shirt with no coat; the shirt sleeves were very wrinkled, I noted.

I looked around and saw my lanky friend, Carl, in the seat across the aisle from me. His legs were spread out under the seat in front of him and he was snoring, taking a deep breath and letting it go with a soft whistle. Seeing him was a relief. I knew I was safe; the train whistle hadn't even woke him up. He must be very tired, I thought.

A nice warm blanket was pulled up around my chest. I felt the train lurch as the brakes were applied with a loud piercing shriek. A loud voice called out "Baker! Baker, Montana!" I still wondered how I had gotten on the train, but I didn't wonder how I was going to get off. I heard Carl's voice say, "Hey, Cappie, I'll help you up. How do you feel? I see you're a bit better. This is where you wanted me to take you, and here we are."

He straightened the seat up a little for me and said, "I think your doctor is going to meet us, Cappie."

"Did you call him, Carl?"

"Yeah, you gave me his card last week when you weren't feeling to good. Do you remember?" he asked me. Before I answered, he continued, as he leaned over me.

"Good thing you did. I needed to call him today, when I saw how poorly you were feeling. He said he would be here to meet our train, isn't that good of him?"

I looked at Carl's worried face, and could tell he was concerned about me.

"Oh yeah Carl, I remember, that was very smart of me to give you his card. Doc Weber is a great doctor and a friend, too. I think I've mentioned this a few dozen times." I smiled.

But I really didn't remember giving him the card. My mind just didn't seem to be too alert. I felt I was still in a daydream, reliving the past.

The next thing I remember was Carl helping me down the steep train steps. I saw Dr. Weber with a wheelchair waiting for me. He pushed it toward the steps and hurried to help Carl get me down carefully. He could tell I was pretty weak.

I saw my old friend was concerned as he helped me into the chair. He grabbed my wrist and with his strong fingers pressed for my pulse.

"How do you feel, Cappie?" he asked me, looking into my eyes with a worried look.

"Oh, just dandy, Doc," I answered, trying to smile up at him.

"Well, I got a room waiting for you in the hospital and I'll do some tests when we get there, so let's go."

SHE WAS WIRED DIFFERENTLY

With that, he began pushing me quickly down the station's platform. I saw a car parked with his wife waiting at the wheel.

Carl was following us. He helped the doctor put me in the back seat and followed me in to sit beside me.

"How you feeling, pal?" he asked.

"OK," I lied, resting my head against the back of the seat.

Doc got in the driver's seat as his wife slid over.

"How is he?" she asked.

Doctor Weber mumbled something.

"We're taking him to the hospital," is all I heard him say.

It seemed like just a few minutes until the car stopped and they were helping me out and into another wheelchair. That was fast, I thought, as they pushed me through the brightly lit doorway.

Before I knew it I was in a white-sheeted bed, with Doctor Weber helping me into a hospital gown. I was very grateful for his help; we had talked about this scenario a few times before—how I could protect my anonymity. I knew he would put me in a private room, and he knew certain people who would keep their mouths shut for him, he had assured me.

He had someone come in and take my blood. They left, and Carl came in to say good-bye.

"So long, Cappie, I've got to get back, but I'll call you." Turning at the door, he waved, gave me a big grin, and was gone.

"Good friend you got there Cappie," Doc Weber said as he leaned over, putting his fingers at the side of my neck. Then he took my blood pressure. Getting a reading, he took it again. I tried to read his face, but he didn't let on what he was reading.

"Good thing you got here when you did. I've ordered some tests, and I think you're safe for the night. I've got a good nurse who knows how to keep her mouth shut. She'll be reporting to me any change in you and I'll get the results of your blood work in the morning. You just rest, my friend." And he left.

Did I tell him thanks, I wondered. Everything seemed to be going either too fast or to slow. The nurse came in and dimmed the lights.

HATTIE

After she left I again slipped into a dream state wondering about my life. I had done this almost my whole life, wondering why I was so different. What had made me like I was, different and defiant? I had keep diaries, putting down all my feelings since I was a very small child. I couldn't talk to anyone about what I was thinking or feeling.

CHAPTER TWO

I heard Doctor Weber's voice; he was speaking softly, almost in my ear. "Cappie, can you hear me? You are safe, in the hospital in Baker. You're in a private room and I have private nurses for you," he whispered. "Seems you had a slight stroke. We are doing blood work and tests. Just rest now. Your friend Carl had to go back to Rhame."

I opened my eyes and looked up at my wonderful friend and doctor; his eyes were full of compassion and worry.

I fought to bring a smile to my lips. I felt a pain somewhere deep down in my right side, and I didn't seem to want to move.

I said, "Thanks Doc. I knew I'd be safe once I got to you! You are truly my best friend in the world. Let me go if it's bad; no heroics, remember," I whispered, closing my eyes to end the conversation. I heard him shuffle out of the room and knew he too was getting old. How lucky I am if he outlives me, I thought.

I closed my eyes and began remembering my life. What an interesting one I've had, I thought. I wish I could live it all over again. No one would believe it, I thought as I smiled to myself. What a secret I've kept all these years and am still keeping. Of course I've had lots of help keeping it. It's been like a big hoax I've played. As I lay there thinking of my life, the thought came that if I lived through this stroke problem I could write the story of my life. Let's see, I mused, how would it begin?

HATTIE

Maybe I can relive it beginning with my earliest memory when I was only about six. I've been in the newspaper business all my life writing stories about other folks; I know I could write a darn good one about my own life. I've hid it for so long; I'm tired of hiding it! Rebellion reared its head. I never thought of writing a story about my life before. I spent my whole life keeping my life a secret. But it would be fun to write and shock everyone, I chuckled. Let's see, how would I begin?

I remember when Fred was born and Papa wanted to move West again. I closed my eyes and could picture the room: Papa walking back and forth with his big boots on, swinging his arms, and talking loudly and excitedly to Mama.

I heard Mama say over and over again soon after Fred was born, "It seems like the whole county wants to move West." And sure enough Papa wanted to move West again. This time to Marion Junction, South Dakota, a place he thought would be a good one in which to settle down. He wanted a newspaper of his own because that was the business he liked much more than the carpentry trade of his father, or the teaching job he had when he met Mama.

"Liz, my dream is having a printing office and a newspaper business in a small town, getting to know all the folks. What do you think of that idea for our lives?" Mama, we knew, said yes to anything Papa wanted to do. He was her hero.

Newspaper work seemed to fit Papa's personality best. He was a very curious person and well educated for the times. He had learned carpentry and had been apprenticed by his Pa to a newspaper man when he was quite young; this allowed him to get an education and a teaching position when he was only nineteen. He and Mama were both better educated than most folks.

Papa insisted that all us children learn to read almost as soon as we could talk. All the education he had learned he was determined to pass on to us. Mama also shared his love of education, which she had gotten from her well-educated English father.

Hattie was remembering her Papa filling her head with history. His passion was the presidents and American history, and he shared all this with us children. He felt we were living in the most wonderful country in the world and compared what America offered ordinary folks, like us, to other ordinary folks in other countries.

"Just think," I remember him telling us, "the government will *give* us land, just give it to people. If they want it, all we have to do is go claim it! Free land!" I remember him shouting; "Nowhere in the world is this happening." We could all feel his excitement at this thought. As we moved West we saw little cabins dotting the landscape each on a section of land. Papa said a section of land was 640 acres, one square mile; he pointed out the little cabins to us as we rode in the back of the wagon with all the bedding and boxes.

Papa explained how folks "proved up" on a claim: "First you need to dig a well for water, build a cistern to catch the rain water," He laughingly would say, "Then pray for rain. Most families then planted a vegetable garden and got a cow, some chickens, and began plowing to put in a crop. All this work made the family self-sufficient and free and independent. Freedom, that's what everyone wanted"

I remembered when we got to Marion Junction; Papa built his newspaper office and put in a printing press while the town was still being laid out. Then after the town was through laying out where the roads were to be built, Papa found that his newspaper office was right in the middle of Main Street.

Everyone had a good laugh about that. Mama told that story over and over again, about the joke the city had played on them. Papa had to move the building back before he could start printing. Everyone helped him do this tricky job.

We stayed in this small town for a little over a year, but soon Papa began getting restless again. We girls remember him talking to Mama about how he didn't believe the town was going to grow much; he wanted to move further west. "Everyone is going West, Liz, some as far as the Pacific Ocean." He said this as he paced back and forth on the second floor of the newspaper building, where we were living. Mama said she couldn't drag us four children to a frontier town until he had gone there first and looked it over.

HATTIE

So off Papa went all the way, this time to the Missouri River and a shantytown called Chamberlain, South Dakota. It was across from the Black Hills and was almost in the middle of the Sioux Indian territory. It was also the jumping-off place for the many people who were headed for California, where gold had been discovered. I think Papa would have liked to be going there.

The train hadn't come into Chamberlain yet, so, Mama and we children had to travel by a horse-drawn wagon with all our possessions piled high in it. We three girls rode among all the boxes and piles of bedding. Mama sat up on the long board seat with the driver holding baby Fred on her lap. It was a long, hot, dusty trip.

The driver stopped to pick up another passenger, a man, who also rode up on the long board seat with the driver and Mama. Soon Mama decided to ride in the wagon with us. She sat right behind the driver so she could hear the two men talk. The driver began talking about the Indians who lived only five miles from Chamberlain. He was telling about the cruel ways of the Sioux. He was warning the man to be careful, as some of the Indians were still very savage.

Mama's face turned frightened as she heard parts of the conversation. We girls got kind of frightened too. Even though we couldn't hear the driver's whole conversation, we knew Mama was hearing most of it. She shushed us with her finger to her lips, so she could better hear what the two men were saying.

Just before we came into Chamberlain the driver stopped at the very top of the hill. We could see the town with the Missouri River flowing behind it. I never forgot this sight. The scene looked down on a lot of shanties and tents, and even with my six-year-old eyes, it didn't look like the towns we had been living in or the ones we had just come through. Here it was like a tent city. I heard Mama say, "Oh, dear!"

We started down a long steep hill, and we saw Papa coming on a horse to meet our wagon. We were all so glad to see him. I hadn't realized how much I had missed him until then. He was so excited to see us; he was waving his hat and calling out "Hello, hello, Lizzie, girls!" You could tell he had missed us as much as we had missed him.

SHE WAS WIRED DIFFERENTLY

"Thank God, you made it safely!" he shouted. I couldn't hear everything he was shouting, but it made me feel good to see our Papa once again. It made me feel warm and safe. I didn't know how it made Mama feel, but she seemed very glad to see her fun-loving, adventurous husband. She had taken off her bonnet. Her auburn hair let loose flew all around her beaming, flushed face. We waved and waved, until he came alongside the wagon. I'll never forget that day.

Our first home was a cave on the side of a hill. Papa was building us a shanty. He was a good builder, but he had little time to build our cabin; he had spent so much time building his printing and newspaper office. He had only odd times to work on our shanty and had been so anxious to bring our family together that he had sent for us even before he had finished the little shanty.

Mama hated the cave and couldn't wait for Papa to finish, so when the shanty only had three sides and the roof they hung a big piece of deer skin on one side and we moved in. It still had a dirt floor.

A big storm came up. Mama and Papa had to get up and hold the big flap of skin across the open area where the fourth wall belonged to keep the rain from coming in. The rain just poured down. Many toads came hopping in and started jumping on our beds. We children had our beds right on the dirt floor. Nellie and Mabel screamed in terror, but I tried to catch them and throw them back out into the black night. They felt very slimy, but I wasn't afraid of them. The very next day, Papa finished the fourth side. He put a long board on hinges in the wall so we could open it up for a window.

I remember some of the first Indians we saw. They came up the Missouri river in their canoes. We learned later that they visited all the newcomers, and were as curious about us as we were about them. We had been warned of their visits but it was still a scare. You could hear them coming with bells on their ankles and scary grinning painted faces. Many were toothless. They wore belts of human skulls around their waists.

After we had lived in Chamberlain for a while, we got a cow so we could have our own milk. One day Mama had baked a cake, which we

HATTIE

were all looking forward to eating. She had it cooling under the big open window along with a large pan of milk. I watched as an Indian buck came inside and motioned for Mama to give them some milk and cake. So Mama, who was still afraid of Indians, cut some pieces of the cake. She put them on plates and passed them around to the bucks and squaws in this Indian party. When Mama handed them the plates they ate the cake quickly, drank the milk from our cups and then the squaws took the plates and cups, put them in their bags, and walked away with our precious plates and cups. Mama was sick about the loss. I had never seen her cry before but she did that day.

Papa was doing a lot of business with the newspaper. His was the first paper in the area; he called it *The Chamberlain Chief*. He built us a nice home from that one-room shanty in a few short years. We watched the little town grow from the tent city to a young frontier town. There were still no railroads, so all our supplies were brought up on big steamboats from the river. When we heard the boat whistle, all the townspeople would stop whatever they were doing and run down to the waterfront to get their supplies and, most importantly, their mail, of course. Everyone wanted to hear the news from the east. Also we all wanted to look over all the new settlers coming to join us. Everyone wanted to see the excitement; even the Indians came for a look.

When the boat whistle blew we knew it would be a fun day. My little brother, Fred, who had stolen Papa's heart, had now stolen mine too. He tagged after me everywhere. I could run fast but he could run almost as fast and far as I could, so he and I would race down to the dock to watch the men unload the goods and witness the new settlers getting off the steamboat. The settlers came to farm the hills near town and prove up on a claim for the free land.

The little tent village was finally beginning to look like a real town. One day the rumor came that the railroad was coming into Chamberlain! The settlers had built a church and a small schoolhouse for us children to attend. Also, there were two more stores built alongside Papa's printing office, and a hotel called Pillagers.

Chester Arthur became our twenty-first president in 1881, about the time our family had moved to Chamberlain; he had opened the government land for settlers. This caused a lot of trouble with the Indians, because, of course, everyone wanted to be near water and in the flatlands. When Grover Cleveland became president in 1885, four years later, the first thing he did was recall the government land to be held until the Indians had first choice. The settlers could farm the hills near town, but up north was still government land, which was not opened for settlers.

Papa bought a picture of President Cleveland and his new bride. The president was married just as soon as he took office. Along with the recall of the government land being kept for the Indians, Papa thought the president getting married in the White House was a very proper thing to do. Many people did not agree with Cleveland, especially about the land being kept for the Indians. Many of the settlers had to give land back that they had been developing for themselves. This made them angry, but Papa thought it was the fair thing to do. "It's fair, give their land back!" read his headline. Papa loved to laugh and poke fun at folks, which made him unpopular with some.

I guess Chamberlain became too busy for Papa, because some people talked him into going down to Wheeler in Charles Mix County to start another newspaper there. It was not far from Chamberlain, right on the Missouri River. Newspapers were very important at this time because people wanting to claim land had to publish the claim in a newspaper to make it legal.

Mama was not so happy about this move, but I was ten years old and I was worried as Fred was getting almost as big as I was; this made me mad. I quizzed Mama on how I could get bigger. I thought she knew everything. I remember saying, "Mama, if I eat more bread, will I get as big as Fred? He eats so much bread, and he is getting as big as me." She said "Hattie, why do you want to be big? Women should be smaller, and you are a perfect size for a girl." That was not the answer I wanted and I remember being very angry. Didn't she understand what I wanted—no, felt—I needed? To be big and tough! This was a big scary country to live in and somehow I knew that you had to be tough

HATTIE

to get respect and I wanted to be respected.

There was no school in Wheeler. This pleased me because I liked it much better when Mama taught us at the kitchen table. I seemed to learn more from her. She would answer all my questions. I was able to ask them more easily than in a school room full of gawky, gangly, laughing boys: Mama seemed to enjoy teaching us as she went about her cooking and washing.

There was a white man married to an Indian woman living close by. They had a girl about my age and two small boys. One of the boys was Fred's age so we did have some friends. We played with these children called half-breeds; they were half Indian and half White. Their father ran the only store in town. The store, painted red, also housed the only saloon.

We learned a lot about Indian culture from their mother. She spoke broken English, but we could soon understand her. I began to have respect for the Indians. She taught us about plants and Indian medicine. She showed us the plants to eat and not to eat, and the plants to use for medicine. This information became useful for Fred and me out on the prairie. We even taught Mama some of the things we learned, and she began to plant medicine plants in her garden.

We hadn't been in Wheeler very long when we heard that the Indians decided to have a war dance near town. Fred and I went with the half-breed children, which allowed us to melt into the Indians crowding around the open dirt field where they were going to do the dance.

The Indians were dressed up with plenty of feathers hanging in long stripes from the crowns of their heads down their backs to their feet. The braves had all sorts of fancy articles, such as bells and feathers, fastened to their feet. Some had belts around their waist hung with scalps all around—some with long hair, some black, some blonde. Some I noticed (much to my horror) were little scalps from babies. They had taken the babies from the white people in time of war, whispered our friends.

The Indians first built a fire in the middle of the big dirt circle. The

SHE WAS WIRED DIFFERENTLY

Indian brave that took part had a long piece of sinew, which had been cut from a buffalo. He drove a stake near the fire. Each Indian in the circle fastened his piece of sinew to the stake, and then took the other end and fastened it to himself through a slit he cut in his own chest. They began to dance to the beat of a tom tom. Blood began running down their bodies. They danced faster and faster, and seemed in some kind of a trance. I was fascinated; some of the young Indian men began to stumble a little. Our friends whispered to Fred and I that if one of the Indian braves fainted or gave up they would be branded as a coward. Then they would be made to leave the tribe forever after this dance and would be made to wear squaw's clothes for the rest of their lives, in disgrace. Their mother had told them this, so we knew it was true. We heard later that this was to be the last war dance, as the government outlawed war dancing.

One day soon after the dance, word came that lots of Indians were coming our way down river in their canoes, dressed up in paint and feathers. Word was sent around that maybe they were on the warpath. The men in town decided that all the women and children should go to the store in town and stay together.

I had been in the store section of the building many times and knew a big board wall divided the saloon from the store part. I had never seen the bar side, but we could hear the music and laughter sometimes while we were shopping. I'm sure neither Mama, nor my sisters, had ever seen the bar part.

Somehow in the fear and excitement we were herded into the saloon. At first it was kind of dark, as there were no windows and only dim lights, but my eyes became accustomed to the soft candle light. I saw two women I had never seen before in town. One was white, very white; her face had been powered with what looked like Mama's white flour and her lips were a bright red. She had deep-set blue eyes, and she had on a beautiful soft dress. I didn't know what it was made of but Mama told us afterward that the beautiful material was silk, and she explained about the Chinese silk worms. Nellie and Mabel couldn't think or speak of anything else for months except how beautiful the dress was.

The second woman was obviously a young half-breed. She was beautiful as only mixed breeds sometimes are. She immediately fascinated me. She had on a dress of the thinnest buckskin. It seemed to me as though it were her skin as it was almost the same color. She was wearing a beautiful turquoise bracelet and necklace. I was in awe of her. She had jet-black hair and black eyes with long, black eyelashes. Those eyes told her story; they were defiant as she looked us over when we burst through the door. She stood up from a stool at the front of the bar where she had been sitting with a rather devilish-looking young skin trader wearing a bright red shirt. He was a "Bull Whacker." Papa had warned us in Chamberlain not to mingle with these rough men.

These Bull Whackers, Papa said, were men who used mules to haul freight into the Black Hills across the river from us. They were a hard bunch. They wore leather pants and usually a heavy plaid shirt, but when they dressed for the evening they all wore bright red shirts and rode their mules up and down the streets to and from the saloon. When we heard them riding and whopping and hollering in Chamberlain, Mama would put us to bed early. It was dangerous to have any lights on as these rough men would shoot out any light right through the window as they rode by, just for the fun of it. In the morning they would pack up their mules, wagons, and freight and cross the Missouri River in big flatbed boats on their way back to the Black Hills.

So when I saw this young man wearing the bright red shirt sitting on the bar stool turning to look at us, I curiously looked him over. I thought maybe I should be afraid of him but for some reason I wasn't.

The half-breed young lady spoke first to us. With a hand on her slim hip, she said, "What can we do for you people?" She spoke in a heavy accent. The old man Papa told to bring us to the store just stared at her, as if he too had never seen her before, and he mumbled something that she seemed not to understand. Mama stepped up and said, "A group of Indians were seen coming down the river toward town. We don't know what they want, but they had their war dance the other day, and we are not sure if they are on the warpath or not."

SHE WAS WIRED DIFFERENTLY

We could see this young Indian girl turning pale when she heard this news, and before we knew it, she had whirled around and run to the back of the saloon and out a back door. We knew right away that she did not want to see these Indians any more than we did, and she was fleeing to a safe place.

I was amazed at my reaction to her. I wanted to run with her; she was the most beautiful person I had ever seen. I wanted to keep looking at her. My heart was in my throat as I watched her fleeing figure. I felt anxious for her; her fear became my fear, and I knew I would never forget her or this moment. I was so young, very close to eleven and I would forever relive this moment in detail.

We heard later that our men marched in a group to greet the Indians in a friendly way. Papa told us afterward that the Indians had decided to come to town to drink some "moon shine," as they called whiskey. That was almost as bad, Mama said, as them being on the warpath, but this time it seemed they did just want to have a good time and they left when they had finished drinking and went back to their camps.

After seeing the half-breed girl, I began daydreaming. One of my dreams was to run with this beautiful girl into the woods; there we would play together, and I named her Doe. She came into my well-developed imagination almost daily from then on. I cherished this encounter. I loved to daydream; I remember I would sit for hours dreaming of "what if." "What if I were a king." "What if I were a deer." I spent many restless nights dreaming, as we three girls slept on a small mattress in the tiny cabin.

Father and Mother had left their nice large home in Chamberlain for this small cabin in the woods. It seemed the whole family became caught up in the day-to-day task of survival. Everyone in the frontier towns, including the children, worked hard from sun up to sun down simply to eat, drink, sleep, and keep warm. Our family was surviving but it was taking its toll. Papa was no longer the happy, joyful young man who had met the wagon coming into Chamberlain just a few years earlier. He was drinking more and more; we all heard Mama complain about it. She was worn out caring for all of us.

HATTIE

I grew restless and went off by myself to read. Books were my friends. I especially liked adventure stories where people went against great odds to do what they felt was right. I read the bible stories with new eyes, asking Mama many questions until she would say, "Enough Hattie, enough questions now, I've got much work to do. You and Fred do your chores now." The history of our country intrigued me most. I re-read the life story of Ben Franklin that Papa had found for me until the pages began to fall apart. Here was a man who did what he wanted and needed to do to save our new nation. He was my hero, and he was also a printer like Papa, whom I still adored in spite of his drinking and long absences recently.

There were no railroads into Wheeler, so the steamboats were our suppliers and we freighted things down from a place called the Bijou Hills. One day Papa heard that a steamboat was coming up the river and was going to stop for fuel. The boat was bringing Sitting Bull, the big Indian Chief. We all went down to the river to shake hands with the famous chief. Mama told us afterward that Sitting Bull was one of the most noted chiefs in the Battle of the Big Horn.

It was hard living in Wheeler after Chamberlain, especially for Mama, and winter came on hard that first year. We were almost out of supplies. Papa and some of the men in town decided that they would walk to the supplier in Bijou Hills, which was the closest place. They decided to take a big bobsled and all the blankets we could spare to wrap the food to keep it from freezing.

We waited and waited for them to come back. When they finally came, it was such a big disappointment. Very few supplies had survived the trip. Papa told us about the problems they had. The big barrel of coal oil they were bringing back for the lanterns had leaked onto several sacks of flour, ruining them. The potatoes were half frozen even after they had wrapped them in the blankets. I saw the sad look on Papa's face while he was telling us this. He looked so tired and his beard was still full of snow. The twinkle was gone from his eyes. We could smell the whiskey on his breath as he kissed us girls. I felt so sorry for him, but I was glad he had gotten back safely. Then I saw the worried look

SHE WAS WIRED DIFFERENTLY

on Mama's face. She really didn't know how we were going to manage for the rest of the winter. I heard her ask Papa if he thought there would be enough food. I didn't hear his answer, I had fallen asleep. I did remember before going to sleep, praying harder than I had ever prayed.

That was a tough winter. We children had to take turns grinding corn in the coffee mill so we could have what Mama called Johnnycakes. We realized how much corn it took to grind just one Johnnycake, so we tried not to eat too many. We ate lots of beans in place of potatoes. I liked beans, so I didn't mind much. I remember that Mama seemed to be getting older, and her hands trembled. I didn't know if it was from the cold, or that she just was not eating enough.

We lived a little over a year in Wheeler and then moved back to Chamberlain, as Mama was expecting another child. I could just barely remember Fred's birth, but in frontier towns you grow up quickly. It seemed that most women were either nursing or expecting a baby and sometimes doing both, so I wasn't surprised or worried about Mama having another baby. But I noticed that both my sisters seemed worried. They whispered a lot together looking at Mama, and rushed to help her lift some heavy pot. They would beg her, "Lie down, Mama, we will do that for you."

I will remember that year in Wheeler. I grew up there. I knew I would never forget Doe, nor would I wish to forget her.

Tintype Fuller Family 1881
Fred, Melroy, Mabel, Hattie (age 6), Nellie, Elizabeth

Picnic Fuller Family 1887 - Mabel, Nellie, Elizabeth, Ethel (baby), Hattie (age 11), Melroy, Elizabeth, Fred

IAPI OAYE.

we Wakantanka wicada keya qa muiakaṡtanpi qa wakantawicuton, eṡa waci, qa wicaṡtapi wicohan kin hena c takuda lica, qa ceon cin kinhan Wakantanka nina iyokipi kte ṡni. Ecin tuwe tinwicakte qa tuwe wamanon kinhan heon yaṡupi ece, wicaṡa ekta hecompi. Tka iyotan Wakantanka wicaṡa owaṡinna tanyan unpi kta cin. Heċen nitakuyepi, tona mnianicaṡtanpi, qa wakantawicuyatoupi eṡa, anpetuwakan kin icunhan ḣtanipi ṡni po, qa wacipi ṡni po. Hena ikceekeya cajeyatapi wadake do. Qa Okodakiciyewakan kin owasin wanna icaḣ aya on cantewicawaṡte ojuna, wicaṡawakan kin hena wicawaka, wicohan kin hena tawapi dakaṡ hececa.

Eya kolapina matakuṡni tka hepa. Wicohan Wakantanka tawa econpi kin ota ecamon, tka ḣtakinipi kin tanyan wakuwa ṡni, tka Wakantanka en micicu he hecetu we do.

BATTISTE DEFOND.

WASINTE.

CANYOWANASAPI, Dec. 25, 1883, hanyetu kin he Waziya hi, qa oiyokipi tauka unhapi ye do. Taṡpantanka koka yamni qa aguyapi suksuta koka wanji, nakun taku-ota ahi. Wicaṡta ṡakpe wohdakapi do. Wicaṡta iyecen ihduhapi kta he cajeyatapi do. Dena wohdakapi. John Tatankaḣota, Adam Wakinyanciqadan, Elias Kahaminyanke, James Akicita, Tatecaḣnajin, qa Titus Icadusmani hena eepi.

LOUIS MAZAWAKINYNANA.

OKODAKICIYE WOTANIN.

OPIICIYAPI.

Wakpaipakṡan Okodakiciye, Jan. kin, wanna wicohan onṡpa on ihduhapi kte. Hecen tokeṡta okodakiciye wanji hen unkaġapi kta nacece.

Oahe Okodakiciye Kiikduṡpapi. Pehanwanmli kaken oyaka. Titankaohe en okodakiciye yanke cin he Jy 13, 1884 anpetu kin he noupakiya kiikduṡpapi. Eya eḣpekiciyapi ee ṡni tuka wicohan noupakiya econpi kta, qa wakantanka oie kin oyate owasin yuhapi kte, qa on nipi kte. Hecen Miniṡoṡe huta wiyolipeyatanhan okodakiciye wanji icaġe kta cinpi. Heon David Lee waḣokonwicakiya kaġapi. Qa tuwe en hunkayapi un kta iyececa odepi unkan niye e econmaṡipi. Hehan nakun wamnayan wanji kaġapi: Thomas Natala hee.

Nakun iṡ Oahe en iyecen econpi. Cetangi wawahokoukiye kin econṡipi; qa hunkayapi kin he Matogleṡka hee, qa wamnaye kin he Matu hee. Dena owasin wi akenom ewicaknakapi.

Hehan miye dehan wicohan wanji nakun ecamon. wauhihan David Lee oyate wanji en waonṡpewicakiye kin hen miṡ waniyetu de ecamon. Anpetu wakan qa anpetu itopa wacekiyapi hena ohinni ecamon kte, qa anpetu izaptan winyan wacekiyapi kin he winyan mitawa econ ece.

Baptizma.

Canyowanasapi en, Sept. 3, 1883 Lizzie Jane he Thomas Katate cinca. Nakun Ewe, James Akicita cinca.

Ihanktonwan Owakpamni, Dec. 16, 1883 Esther Tiyowaṡtewin, Sunkawakuwa tawicu icicons muiakaṡtanpi. Qa Jan. 6, 1884 winyan topa icicons muiawiċakaṡtaupi: Susanna Wicnut he Tiyowaṡtewin okodakiciye tecanien opapi. Qa Jan. 6 qa 13 anpetu henans en dena nakun tecanien opapi: Rebecca Uncagewin, Suzanna wicupi, Emma Tiċnupiwin, Ada Hinyanjicewin, Anna Necklace, Gay williamson, Thomas C. williamson, Horace Caṅhdeṡka, Isaac Mandan, qa Pejutawin, Qa dena titokan kdipi: Robert Clarkson Iḣta, John Iṡtamaza, Peter La Point, Moses Cetanwakinyan, Thomas Ṡtow Maḣpiyaḣota, qa Mrs. Ellen La Point.

Miniwakan Mdecan Okodakiciye dena teca opapi. Dec. 9, 1883, Jane Tateyuḣewin qa Sarah winyanhotewin; qa Dec. 23, 1883, Julia winona; qa Jan. 6, 1884 Charles Thompson Hepan, qa James Ross Matonpaġa.

Canyowanasapi, Dec. 30, 1883, Solomon Piyaiyotanke, Shepherd waṡteyanajin, Susan Mniezawin, hena teca opapi.

Tawacintoaṡte, Jan. 27, 1884, nom teca opapi, Mr. M. P. Phillips (he waṡicum), qa Thomas Crawford, Jr.

Santee Agency, Neb. Jan. 13, 1884 dena Ohuiḣde Okodakiciye en opewicakiyapi: Joseph Powell Maḣpiyacanhdeṡka, Frank Pattinaude, Lucy Lovejoy, Eugenia La Moure, Bessie Barker, henn icicons opapi. Qa hehan dena tokantanhan wowapi ahipi qa en opapi: Mr. william J. Phillips tawicu kici (he Ateyapi wowapi kaġe cin hee), qa Miss Jennie E. Kennedy.

WICAṠTE.

IHANKTONWAN PAHATA, Dec. 20, ciye niua iyokiṡicapi do. Tuka Wakantanka onṡiundapi kinhan okodakiciye kin de wacintanka wanji icaġe kta. LOUIS MAZAWAKINYANNA.

KINWICAṠA TIPI en Dec. 22, 1883 Ojinjintkaṡa he ṭa. He Cincopina cinca hokṡina waniyetu ṡakpe. Isaac Tutehdoka he oyaka.

IHANKTONWAN OWAKPAMNI en, Dec. 25, 1883, Victoria Williams ṭe he wicicama waniyetu yamni, hunku Helen Aungie Williams eciyapi, hiknatou ṡni chan den waonṡpekiye, unkan dehan cinca yamni, unkan tokapa he ṭa qa cante ṡica, tka Wakantanka wacinyan. Nakun anpetu kin he hokṡina wan caḣhukun iyaya qa icimana tuwena wanyake ṡni. John Morgan eciyapi, atkuku Heliakawaubdi eciyapi, qa tuukanṡitku Mawacepa eciyapi. Caġa suta ṡni tka akan hokṡina ṡkatapi qa hececa. He Jesus tonpi anpetu qa uukiyuṡkinpi unkan sanpa Wakantanka onspeunkiyapi.

Hehan Dec. 31, 1883, Wayawa tipi en hokṡina wan Paul Rhodes eciyapi ṭe, he Psicawakinyan takojakpaku.

Waḣicinca wakpa etanhan Robert Hopkins Caskedan deċen woyake. Nakaha Dec. 8, 1883 wincinyanna waṡte makite ye do. Hanyetu he huuku cinca em wanka, unkan ihnuhan hunku he ceya niyan nawaḣon, hecen ekta wai. Unkan wanna ṭe kta iyehantu qehan, kohanna kicison wo, epa. Unkan kohanna econ, qa kiciyuṡtan, hehan ṭa iyaya. Hecen awacin manka. Unkan ito canohnaka wanji we-

"Grim" Fuller Family 1890
Fred, Hattie (age 14), Nellie, Mabel, Elizabeth, (seated) Elizabeth, Ethel, Melroy

Salem, South Dakota 1891

Husband: Melroy Albert Fuller

Born: August 25, 1851 in: Buffalo, N. Y.
Married: September 13, 1871 in:
Died: March 28, 1932 in: Bakersfield, Ca.
Father: David M. Fuller
Mother: Roxana Barnhart
Other Spouses:

Wife: Elizabeth Jane Taylor

Born: February 12, 1851 in: Merthyr Tydvil Glenmarganchire, South Wales, England
Died: November 02, 1903 in: Spring Valley, Minn.
Father: John W. Gibson Taylor
Mother: Mary Lake
Other Spouses:

	CHILDREN	
1 F	Name: Nellie Louise Fuller Born: July 13, 1872 Married: July 27, 1892 Died: September 16, 1953 Spouse: Daniel Franklin Carey	in: Spring Vallley, Minn in: Salem S. D. in: Los Angeles, California
2 F	Name: Mabel Fuller Born: June 16, 1874 Married: June 1892 Died: Spouse: Emory Knotts	in: Sleepyeye, MN in: Salem S. D. in:
3 F	Name: Hattie Fuller Born: March 30, 1876 Married: Died: December 11, 1946 Spouse:	in: Shiefield, Iowa in: in: Baker, Montana
4 M	Name: Frederick Forest Fuller Born: February 27, 1880 Married: Died: November 15, 1952 Spouse: Nellie Ursula Rice	in: Mason City, Cerro Gordo Co. Iowa in: in: San Bernadino, CA
5 F	Name: Elizabeth Fuller Born: 1882 Married: Died: 1950 Spouse: Jack Lyste	in: Chamberlain, S.D. in: in: Los Angeles, California
6 F	Name: Ethyl Fuller Born: 1886 Married: Died: 1978 Spouse: Chris Rygh	in: Lower Brule, Indian Reservation, near Chamberlain, S. D. in: in: Los Angeles, California

Melroy Fuller Family Page

CHAPTER THREE

Back in Chamberlain another baby was born. Now we were four girls and one boy. I don't remember if Papa was sorry that he didn't get another son, but Mama was happy. She laughed and said, "Well, we'll have to save all the girls' hand-me-downs for one more." They named the new baby Elizabeth, after Mama.

In those times, having five children and a husband who moved around chasing his dream was very hard on a mother. I was beginning to see how difficult a woman's life was. It seemed women worked from morning to night, seldom going anywhere except church. Men, on the other hand, worked hard in the daytime but always seemed to have time to socialize with other men in town. I felt sorry for Mama even as I was beginning to look at her life with contempt, thinking somehow it was her fault.

Papa had taken out a claim on some land northeast of Chamberlain. He had built a shanty on it and decided to move the family out there so he could "prove up" on it, because if he didn't improve the property, he would lose his claim. "Everyone is land crazy," he told us, "If I prove up on this claim, I'll own a section that's 640 acres of prime land. I'm sure I can sell it for a good profit. New settlers are coming into town every day looking for land."

Mama didn't want to go. She was still nursing Elizabeth, but, poor dear, she finally agreed. We all loved our papa, and we trusted him to take care of us, so we moved once again.

The shanty still had no floor in it, just walls and a roof. Mama didn't feel very well, so we children did most of the work. Fred and I carried water and brought in the wood, while Nellie and Mabel did the cooking and cleaning, and took care of baby Elizabeth most of the time too. It took Fred and me the better part of each day to do our chores, getting up when it was still dark, lighting the wood stove, and then washing ourselves with ice-cold water.

Mama came down with pneumonia, so Papa eventually had to move us all back to Chamberlain. We children were very happy as we had grown to love the town. It was wonderful to be in a real house with floors again and where we felt safe. We were all worried about Mama being so sick.

One day, Papa decided to go out to his claim and check on the shanty. He thought he might have a buyer for it. He took Mabel and me with him. I had reason to remember this day well. The shanty was several miles from Chamberlain. We had brought our lunch, planning to be gone the whole day. After we ate lunch, Papa, stepped outside the door; he came back in quickly and shouted, "Girls there's a prairie fire coming our way!" We had heard talk about these devastating fires, but we had never seen one. I could tell Papa was very worried. He began yelling and rushing around, "Look girls! Look for something to protect you from the smoke! Find something to cover your heads with."

We looked through all the boxes stored in a corner of the shanty. Finally Papa came up with some old potato sacks and throwing them to us, said "Girls, if the smoke begins to come into the shanty, wet these sacks and put them over your heads .I'm going out and build a backfire to meet the prairie fire. Then I'll go and help the men who are fighting the fire from the other side. You girls stay here!" All the while he was shouting this information to us, he was frantically shoving the boxes around, looking quickly into each one. He grabbed a shovel, stuffing matches and a big handkerchief in his pockets. His last instruction was so fierce, we knew he meant it. "You girls stay here!" He shouted once

SHE WAS WIRED DIFFERENTLY

again as he ran out the door. We had never heard him give an order to us in this tone of voice before, so we knew we were to stay put.

He rushed out the door. We were alone. Mabel and I ran into each other's arms, hugging each other for comfort. With Papa gone we were really frightened. We watched throughout the afternoon as the fire swept around the shanty. We soon lost sight of Papa in the smoke. Mabel and I kept hugged each other, holding tight onto the potato sacks, wetting them down frequently, as the wind blew smoke into the shack. We prayed and watched for Papa to come back.

Mabel and I remained there all afternoon, safe because of the backfire our smart Papa had built. Finally he came back. I gasped when I saw him; he was covered with soot, and looked like a different person. "I think I've aged 10 years today, girls," he said, "but we got it out. All the men worked their hardest. We had a bunch of the best hands in town helping, and lots of Indians came to help, too, thank God! I could hear you girls praying for us," he said, putting his dirty, soot-covered arms around us. "Let's go home."

We were all exhausted by the time we arrived home. I didn't even remember him telling Mama what had happened to us that day. Mabel and I were so tired we fell into bed. The next day everyone came and congratulated Papa, on helping put the fire out. The horror of prairie fires was stamped on my memory.

Our stay in Chamberlain was again brief. Papa heard that bids were being taken for a job building a Catholic boarding school and church on the Crow Creek Reservation, about twenty miles north of Chamberlain. He put in his bid and was chosen for the project. He seemed happy. This was a big job that would take him some time to build.

We knew Mama was pretty tired of all this moving about. She seemed especially disappointed this time. Fred and I carried boxes to the wagon. Mabel was up in the wagon placing each box carefully along the walls. Mama ran back and forth with instructions of exactly where each box was to go. Papa brought the big barrels to hold the heavy household items in. This was usually a chatty, talking, laughing, fun time, while we were packing our belongings in the big wagon. But I heard Mama grumbling for the first time: "Melroy, how many lanterns will we need?" she questioned, stopping in the middle of her many trips to wipe her brow on her handkerchief before tucking it back into the sleeve of her worn work dress.

"Probably only two or three; this place is very small," he answered her.

"I suppose it is filthy to," she said. "I'd better bring the makings for plenty of soap; girls, find the cans of lye we have stored. We better take them all." She said this in a weary voice.

I began to think that the life Papa found so "adventurous" was very hard work for Mama. By this time, we had lived in Chamberlain long enough for Mama to have made some lady friends. My sisters, Fred, and I were beginning to enjoy the peaceful routine of school and church with other children our age to socialize with. Now, once again, life was going to be harsh for the family. I pretended not to notice Mama's obvious sad mood, as she picked up things then set them down again, sighing, "No room for this," and saying, "Girls you may take your diaries but only one book each."

With our wagon loaded with supplies, us children in the back, we headed north for Crow Creek. It was all open prairie land, barren sandy earth, and scrub brush. When we arrived at the place where the church was to be built, we found one lone shanty. We were all shocked to think we had left our lovely home in Chamberlain to live in this poor little dwelling!

Mama grumbled to Papa, "This reminds me of Chamberlain when we first arrived, only worse. Even Wheeler was more developed than this place!" We children were disturbed since Mama had never before seemed so unhappy, or so worried. It made us all unhappy too. Papa needed her help to feed the men he would be bringing in to help him with the building. I saw him put his arm around her shoulder a few times, but Mama just shunned his efforts to make peace. They had always been equal partners in feeding and caring for our family, but the partnership was not a happy one on this day, we all noted.

SHE WAS WIRED DIFFERENTLY

The only other person when we arrived was a Catholic priest, Father Willard, who was from Wisconsin. While waiting for the church and school to be built, Father Willard lived in a big tent and did his own cooking and washing. He conducted church services and school lessons in his tent for the Indian children.

Father Willard was anxious for Papa to begin building. He said, "These Indian children are being lost to heathenism, ignorance, and neglect. It is our duty that they be taught the word of Jesus and also learn how to read and write." He explained, "When the church is built, we will be able to give them an education and save their souls as well." He was so emphatic and serious about his mission for the Indian children that Mama took a liking to him right away.

Not long after we arrived, I saw Mama walking down the path to the creek with Father Willard. He seemed to be talking earnestly to her and she came back with a calm look on her face instead of the unhappy look that she had worn for so many days. She told us children that Father Willard had explained how important the work was that Papa was doing for our Lord and these unfortunate children. He said that we were all working for the Lord. It made Mama feel proud, so we pitched in to help. She seemed happy after this. We all became happy once again. Somehow, though, I couldn't quite forget the desperate look on Mama's face during those first days and hoped it would not come back.

A crew of men soon arrived to help Papa lay the foundation for the church. It was made of big stones that they hauled from all over the prairie hills to the building site. After that came carpenters who built the walls. The workers lived in their own tents and made their own breakfasts. Mama fixed big dinners and packed big lunches for them. Of course Nellie and Mabel worked right alongside Mama, who was still nursing Elizabeth. I again felt sorry for her; the work was so hard.

Nellie took over a lot of the cooking that Mama usually did. Mabel, Fred, and I took our orders from her as well as from Mama. None of us complained, because we all wanted to help. I began to notice that Papa stayed later and later at the worksite, even after the men had left for the day. Sometimes Fred and I would go to call him for supper; we could tell he had been drinking. He laughed a lot and made silly little jokes. Sometimes he came back with us, but many times he stayed way past supper time, and Mama would set his plate on the stove to keep it warm. I tried to remember that we were helping the Indian children like Father Willard said and that this should make us all happy.

It seemed all the workers did was work very hard, sleep, and eat. They went to town occasionally, on payday. Mama complained to Papa that when they came back they were very loud and used unruly language, "unfit for children's ears." she said. We heard her complain very loudly when one of them brought a woman back with him one night. We all heard Mama talking to Papa loud enough for us to hear her say, "I will not have these goings on." She insisted Papa speak to the fellow immediately, and he did.

Even though her talks with the "Good Father," as she called Father Willard, were helping dispel her unhappiness, a new problem arose. Mama hated drinking and we all knew Papa had taken up drinking again. When we were in Chamberlain he had almost stopped entirely.

It was lonely for Mama, with no families near us like there were in Wheeler and Chamberlain. Nellie and Mabel didn't like living there either, but Fred and I loved it. We had fun playing near Crow Creek, not far from our shanty. We caught pollywogs and watched them turn into frogs. We caught a fish or two for dinner and received high praise from Mama. I remember this as a happy, leisurely, fun-filled summer. We did our share of the work, but there was plenty of time for exploring.

I was beginning to develop a stronger opinion about the differences between men and women's roles in society. It seemed to me that Papa had more fun and free time. I knew the construction work was very hard, but all the men seemed to enjoy what they were doing. The comradeship between the men was apparent. We would hear them joking and laughing at one another over something that we never could figure out, even when we stood watching them at work.

Mother, Nellie, and Mabel's work seemed like thankless, repetitious work, day after day the same thing. It seemed boring, I wrote in my diary. "I hate women's work. Today Nellie asked me to peel several onions, and they made me cry while I was peeling them, so when I couldn't see, I cut my finger. Nellie grabbed the knife from me, called me clumsy, and said, 'Go tear a piece of rag off that old petticoat in the rag bag and tie your finger up. Then bring in some more firewood. That you can do well, Hattie, but in the kitchen you are of no help at all.'" I wrote in my diary, "I'm ashamed."

I sometimes talked to Fred about this. I remember telling him "Too bad, Fred, that people have to eat. Think of all the time Mama, Nellie, and Mabel spend cooking for everyone. It takes them all day and is eaten in such a short time, and then, the dishes have to be washed and put away. I saw that fat man, Edwin, last night. Nellie served his plate, and before she got the next fellow's plate ready, he had gulped his food down and wanted more!"

Fred and I were sitting on our favorite rocks with our bare feet dangling in a small stream of cold water. He was splashing the water hard so it would fly up and spray him. He said, "Yeah, I have seen him eat like that, like those Indian dogs. They would bite your hand off if you tried to take their food away." He laughed, "I would hate to have to cook all the time. Nellie likes it, but Mabel doesn't. I heard them talking and Mabel said she wants to be a writer." I thought about Fred's remark for a few minutes then said quietly, almost to myself, "I want to do what Papa did when he published the Chamberlain newspaper."

Fred looked at me for a moment, then said, "It would be very hard for a woman to run a newspaper."

"I will find a way." I said stubbornly, "Maybe when Papa is done here he will start up another newspaper and I can offer to help him."

Instead of laughing at my idea, Fred looked thoughtful and said, "I think I will offer to help him too. Maybe he would let us help interview people!"

Then I had a sobering thought: "Well, first Papa has to start another newspaper. You know he sold the Chamberlain Chief, and there isn't room for two papers in town." Fred agreed and we dropped the conversation, but I remembered this talk because I finally understood what I wanted to be, a newspaper owner. I wanted to live a man's life.

HATTIE

I hated a woman's life—having children, cooking, waiting on others. I wanted to be like Papa.

Across the creek lived a half-breed Indian woman who had a son about a year older than me, which made him about thirteen. His name was Half Moon. He told us that the half moon in the sky was the first thing his mother had seen after he was born. The shanty they lived in was very dirty, but he seemed to be clean so Mama allowed him to visit with us. We were all lonely so we welcomed a new friend. As the son of a half-breed woman, he was looked down on by his Sioux tribe. But despite this attitude, the Indian men had taught him many of their ways; they were very different from ours.

I admired Half Moon. He did not seem beaten down by living on the reservation like most Indians seemed to be. I understood that his people had roamed free over the prairies for centuries, but he was too young, Mama said, to remember those days.

I looked into his black eyes narrowing into slits of anger when he talked about the white men killing the buffalo, not for food, or hides, but for the land they were roaming on. Those eyes would get soft and misty when he talked about his little brother who was sickly.

Half Moon's father being white had given Half-Moon's skin a light copper tone. For a fourteen-year-old youth, he was well developed, not quite six feet, tall, with broad shoulders that were not fully developed into the muscular strength they would soon become. His black hair had never been cut, so it hung down his back into a heavy braid. I had watched his Mother braid it for him as he sat in the sun outside the large hut his family lived in.

I learned they believed everything was spiritual. Mama called it superstitious. Half-Moon believed that the Great Spirit watched over everything, that if you did not treat the land or animals right, the Great Spirit would punish not only you, but everyone in your family and in your tribe.

I explained to Half Moon that I understood what he meant about a Great Spirit. I told him about our belief in God and the Son of God, Jesus. He listened but was puzzled. "I never see you dance for these

SHE WAS WIRED DIFFERENTLY

things" he said. I explained that we go to church to pray to God, not to dance.

Half Moon began to accompany Fred and me when we went fishing or played by the creek. He seemed lonely. Soon we were roaming the prairie with him as our guide. He taught us valuable lessons that he had learned from his Sioux tribe.

One day he took us to a place he called "prairie dog towns." There were little mounds of dirt as far as the eye could see. Prairie dogs had dug their holes into the ground, leaving little mounds of dirt around the top of the hole. When the animals saw anything large, such as a person or a big animal coming, they would run up on these little piles of dirt, stand on their back legs, and bark. They would then scamper down into their holes.

Fred and I wanted to chase them at first to make them bark at us, but Half Moon warned us that rattlesnakes lived with these prairie dogs, and told us to be very careful. He wasn't afraid of the snakes himself. In fact, he showed us how he could catch the rattlers and bring them home for his mother. She would skin them, cook the meat for food, and dry the skins to use for decorations.

I had no interest in catching them myself and couldn't imagine eating snake meat. We watched Half Moon catch some of the rattlesnakes. He would wait quietly beside a rock with a rope on a long stick. Then he would quickly move a rock with his foot, and watch as a rattler slid out. "Quick as a wink," I told Mama. "He lassoed the head of the snake and by jerking on the rope it choked the snake. Then he lifted his foot up and slammed it down hard on the rope. The rope almost cut some of the heads off!" I explained to her excitedly. He then held the snake up in the air and dangled it so we could watch it struggle; its body whipped back and forth until it died. I told Mama that Fred and I had watched while he killed a number of snakes like that. "It was fascinating, Mama."

Mama was horrified. She told Papa that night that Half Moon was showing Fred and me how he killed rattlers, and she said, "I think it is very dangerous, and I don't like to think Fred and Hattie are learning

HATTIE

how to torture animals like this rather than to kill them cleanly." But Papa replied, "It's good for them to learn these things, Liz, and the young boy is teaching them to be careful, not to torture, it sounds like. It won't hurt to find out what else Fred and Hattie are learning from the boy, but as long as they are learning how to live out in this wilderness, I have no objections." So, with Papa's blessing, we continued to accompany our friend on some great adventures.

Fred liked to hang around when Half Moon's family was eating. He did not have my concerns about what the meals were made of, or how clean his mother was. As mothers everywhere seemed to do, Half Moon's just fed Fred along with her own family and offered him all he wanted. I asked him once what he had eaten and how it tasted, for I knew the Sioux ate dogs. Fred just shrugged and said he thought it was some kind of squirrel; he wasn't sure, but it tasted good to him, different from the food Mama made, he said. Fred was always hungry. He said there were wild onions, roots, and other foods he didn't recognize cooked with the squirrel-like meat in a sort of stew. He said rattlesnake tasted good; it has a bird flavor.

I turned twelve while living in the shanty on the Crow Creek Reservation, and Fred turned eight; he was as big as me. It was then that two very important things happened. The first was going through the physical changes that make women able to have babies. When they first started, I was mortified, thinking I had somehow contracted some evil woman's disease. When I finally worked up the courage to talk to Mama about this unmentionable problem, she gave me a huge smile, laughed, and said, "Nothing bad is happening to you, Hattie. All women do that when they become mature enough to have babies. It can cause some women discomfort, and so if you feel tired or uncomfortable, you are welcome to take a rest." She talked for a while, helping me to calm down and understand.

After our talk, I went to bed and lay there thinking. Mama had said it was natural and not a bad thing. But from what I had seen of life, once a woman had babies, her life became one long series of workdays, one after the other, nonstop, and one baby after another. So I didn't think this was such a good thing. Thoughts swirled through my head as I finally drifted off to sleep. When I awoke it was almost suppertime and the only thought in my mind was food!

By the time Elizabeth turned four, Mama started having Fred and me take Elizabeth with us so she could get a few chores done without her underfoot. Mama would ask us to look for wild fruit so she could make pies and jam. Although the prairies didn't have a lot of vegetation, Half Moon had showed us lots of edible food that grew all along the little stream banks and in the gullies. We picked berries, some of which we knew only by their Indian name. There were also wild plums, currants, cherries, and grapes. What Mama couldn't cook up right away, she canned for mid-winter when a taste of summer fruit was a wonderful treat. Nothing was wasted.

Sometimes Half Moon would spend the whole day showing us where along the creek we could find other food such as wild potatoes, onions, and turnips that grew on vines. We took everything we could carry back to Mama, and she surprised us by making delicious vegetable pies. Many times Half Moon was invited to eat with us when he was there at mealtime. He didn't understand about using utensils or about table manners and just gobbled the food down as fast as he could. Mama said that she had never seen such bad manners, but I knew she was secretly very pleased because Half Moon would praise her the whole time he was eating, saying, "Good, good," over and over.

One day Half Moon came to get Fred and me, but Fred was sick and had to stay in bed. I asked if I could go, even if Fred couldn't. Since Mama was accustomed to us spending the day with our friend, she consented. The church Papa had been working on for all these long months was almost finished. All the men were gone, and Papa was just finishing up on some benches for the church and school. Mama had begun packing our things and Nellie and Mabel were busy helping her. With no crowd of men to cook for, the older girls had time to watch Elizabeth. This meant I could have a whole day free to myself.

Half Moon and I went out to explore the prairie. We had been exploring like this for months. Half Moon had brought his mothers little

HATTIE

pony and we both climbed onto its back, carrying the lunch basket Mama had packed for us.

Soon I noticed he was taking us to some hills that I had never been to before. He said in his broken English, "Hattie, we go to special Indian place, a place Sioux people have gone for many, many years, yes?"

"All right," I remember answering, "I would like to go there!" I admired him; he was a hero to me, a defender of his people, like the strong men I had read about in books. He seemed to know so many things, and he treated me as a best friend, not like other young boys had treated me. The town boys told me to go away; they didn't like playing with "silly girls," they had said to me. I never had a best friend before, and I wanted to be like Half Moon. I knew I was going to miss him and it wasn't likely I would see him again.

After riding for some time, we came to a trail at the foot of some hills. The little pony seemed to know the way and trotted briskly along the well-worn path. On the hillside I saw cave entrances with Indian markings above or beside the openings. I was fascinated. Pointing to the markings, I said in Half Moon's ear, "What do those markings say?" He answered, the wind carrying his words back to me, "Different meanings; some say only men enter, some say only women enter. This place is where we Indian's come to honor the Great Spirit." We passed five or six caves and came to one where Half Moon pulled the pony to a stop. He got down and helped me down.

He tied the pony to a big rock and I followed him, squeezing through the small opening of the cave. When we finally were able to stand up, I found myself in a large circular place almost as big as our shanty. It was astoundingly beautiful. The ceiling was open to the sky high up above us. As I looked around, I saw the walls were all painted in brilliant colors with black-and-white backgrounds.

We sat down on a big rock in the middle of the room. It had a large blanket of buffalo hide on it. Half Moon told me that all of the designs on the walls told a story and he began telling the stories. I could see that this place had an overwhelming effect on him. I tried to follow

SHE WAS WIRED DIFFERENTLY

what he was telling me about the different scenes, but it was difficult to understand their significance and why these pictures were so important to him. I felt privileged that he wanted to share something so obviously important.

The painting on the wall nearest the door opening showed women putting up and taking down teepees. "This shows squaw's work," he said, waving his arm at the first painting. It showed women working together making clothing, cooking, and gathering firewood. The second picture showed women hauling water and wood, preparing buffalo, taking care of the horses and dogs, having children, and caring for the children. The last picture showed the women taking care of a Sioux warrior. All the pictures were done in great detail. The women in the first scene were young, then middle aged and finally as old women. All of the women were shown taking care of the men of the tribe. Here in his culture also, it seemed to me that women performed the drudgery and chores and took care of the men, just as I saw in our life.

The next set of paintings confirmed my observations about the life of a man. Half Moon gestured toward a painting that showed horses, men, and buffalo. He said, "This shows the Sioux warrior before the white man came. This next picture is of a great buffalo hunt with warriors." The next section showed them removing the buffalo's head and using it in some kind of a ceremony. The hides were being made into tents and clothing, and the tails looked as though they were being used as whips and ropes. Half Moon told me they even used the hooves to make what sounded like a glue to attach feathers onto the Sioux arrows. "All these things we learned from our ancestors," he said with pride.

He had become so excited and involved in the telling of the stories that he occasionally jumped up and ran to the next picture. "See this is a medicine man; this one is a chief." He gave names to these figures and then looked at me to see if I understood and if I showed proper respect, which I tried to do.

The rest of the paintings showed Sioux warriors from the past. "This one," he said pointing to a very tall warrior, "Had two wives.

HATTIE

A warrior is allowed more than one wife; it has always been our way. A warrior must be brave, must hunt for meat to feed the people. He must provide animal skins enough to make a shelter and for his wives to make robes. A warrior helps protect his tribe against attacks from the Crow and Blackfoot. It was honorable to take the 'tow ka,' the other tribes' horses when the raiding parties came." Looking at the paintings I felt the people in them were talking to me, telling me about their lives.

Eventually Half Moon went outside to get our lunch basket. We sat on the rock and ate looking at the walls, with Half Moon looking at me strangely several times. We shared the water, and then he took a long cigarette out of his pocket. I had seen other Indians smoking such a cigarette around their village. They also smoked pipes that they passed around when they had special celebrations. I had never seen Half Moon smoke either one before. To my astonishment he lit the end of the cigarette and took some long breaths from it. I watched him with fascination. Papa smoked a pipe once in a while, but the smoke smelled different from Half Moon's cigarette. He then handed it to me. "I've never smoked before," I said, "and I don't think Mama would like me doing it. She thinks it is an evil thing to do just like drinking."

Half Moon urged me, "Try it Hattie, it will make you feel good and the paintings will come alive." This promise intrigued me and I also wanted to please him for some reason I didn't quite understand.

I took the cigarette, held it carefully, put it to my lips, and sucked really hard as I had seen him do. The taste of the leaf smoke on my lips was bitter, and when the smoke went into my lungs, I couldn't believe how harsh it was. I began to choke and sputter, coughing again and again. Half Moon laughed and said, "That's how I was at first, but you will not have so much trouble the second time." I tried to hand it back to him, but he said, "No, little one, you have come this far with me; now we are one, you and I. You must take another breath for me and I will take one with you." I didn't want to, but I wanted so much to please him and be his friend, I once again took a breath. It was true what he said; that breath didn't seem as harsh as the first. We both had a few more breaths from the cigarette, and then he killed the fire on the end of it and put it back into his pocket.

He said, "Let's take a nap here on the rock." Feeling tired suddenly, I lay down beside him and looked at the pictures on the walls. He was right again! The pictures did seem to come alive for me; I was sure I saw the Indians in the pictures moving like living people, among the trees. I turned my head around and around, looking at the different scenes, until my head started swimming. I felt a desire to giggle. Turning my head toward Half Moon, I saw that he had pushed himself partway up, and was resting on one elbow looking down at me with that strange look again. "The smoke is fire water," he said. I didn't know what he meant, because fire water was the name the Indians used for whiskey. Then he did something that so surprised me, I didn't believe he had done it until it was over. He leaned over and kissed me hard on my mouth! I looked into his dark eyes and suddenly felt a desire to kiss him back.

Soon, he had taken my clothes off as well as his own. We were totally naked, lying on that big soft buffalo fur. I had seen him half naked before when we went swimming in the stream with Fred. Boys could get away with taking their clothes off to swim. But I knew this was something very different. Emotions seemed to race through my swirling head, a tremendous desire to run as well as an equally intense desire to stay there. Something had happened to my head. It was saying "Hattie, get out of here! Quick!" But my legs wouldn't move. I wanted him to like me and not be disappointed. So, I lay there with my head swimming from the harsh smoke and the excitement of knowing my "hero" wanted me to stay where I was and let him do what he wanted. I felt his hands touching my body all over, and then I felt the weight of his body on top on mine. Some part of me still felt it was wrong, but another part was very curious and detached, as though it were happening to someone else. I knew from Mama and my Bible reading that I was doing something terribly wrong, but I didn't care right then. He withdrew from me and I felt something sticky and damp trickling in my private place.

HATTIE

When we left soon afterward, my head was still swimming. I could barely look at Half Moon during the long ride back home. By the time we got back, my head had cleared a little with the fresh air. He helped me off the pony and handed me the empty lunch basket. He then leaned close to my ear and whispered in a low, hushed voice, "Hattie, you are now my woman. You understand you are now mine?" I looked down at the ground and said "yes" very quietly. I didn't know what he meant. I was bewildered.

When I went inside, I kept my face down and told Mama that I didn't feel well. She looked at me strangely. I went to bed and lay there, reliving the afternoon in the cave over and over again. I took my diary from under the mattress and wrote, "Today was an important day," and I dared not write any more as someone might read my diary. So I decided to bottle up my feelings knowing I would always remember this day in detail; I didn't need to write about it.

Half Moon came the next day, and Mama told him Fred could go with him but I was ill. I had told Mama I didn't want to go. He came day after day to ask if I could go with him, but Mama told him no. She sensed something was wrong, but she didn't pry and told him that I had to help get things packed for our return journey to Chamberlain. It was true, we were all busy packing. Soon he stopped coming. I saw him once or twice from a distance down by the creek, but never again did we speak to each other. Mama knew something had happened that day, but she never questioned me about it.

It was strange to me about Half Moon. I thought about him often, feeling as though what happened was to someone else and I had been merely an observer. I knew one thing. I would never again idolize any man like that. Somehow I didn't feel bad; I wasn't angry, sad, or hurt. I just never wanted to experience that again. I was glad in a way that it had happen because now I knew more about a woman's body. I didn't like the thought of giving control of my life and my body to someone else. Hadn't Papa always told me how smart I was?

Papa had finished with the school and church. Mama, with the help of Nellie, Mabel and me, soon had the packing done. She looked pleased, saying that it was the quickest job of packing they had ever done. The Catholic Church sent new priests, along with nuns who were to serve as nurses and teachers. They opened the boarding school for all the Indian children the day before we left. The children came in groups and the place was called St. Stephan.

Papa made the big wagon ready for the trip. We loaded everything and set out for home with smiles and laughter. No more shanty living! We looked forward to seeing our friends and neighbors and being in a new school! Chamberlain seemed like a big, bustling city after the solitude of the Crow Creek. We settled in, hoping we were home for good. I knew Mama was very happy because she hummed as she cleaned and cooked. She was loving and kind, never having a mean word to say even when one of us did something awful like sassing her back or taking a bigger helping of food than we could eat. She spoke softly and explained kindly what we had done wrong. She was so special. I knew other mothers who were cruel and some who seemed mean, always whipping their children with switches from trees.

We had barely gotten settled in when Papa was approached about a job at the Indian agency on the Lower Brule Reservation to teach the Indian men the carpentry trade. This move was an opportunity for Papa to get paid by the government to teach his trade to the young Indian men, helping them get off the reservations and make a living.

"This was the hope of our government," Papa said. "We will be killing two birds with one stone. First, we will be acknowledging that we owe the Indians something for taking their land, the land that had supported them with food and survival." We had built train tracks and roads using their land. We had confined the Indians to reservations, and big as most of these reservations were they could not support all of them or provide their old way of life. Our government was trying to make it up to the Indians by teaching them new ways to support themselves. This was the rationale behind the reservation Papa explained this to us children. He was happy to get this posi-

HATTIE

tion because our family would be more secured having steady money coming in every month. We would all be happier and I could tell Mama agreed with him. She began packing, this time with a smile on her face. It was a different Mama than when we were packing for the Crow Creek Reservation where Papa had just finished building the Catholic Church.

CHAPTER FOUR

Living on the Lower Brule Reservation Agency was a completely different experience for our family than the primitive conditions at Crow Creek.

There were two farmers, a blacksmith, and Papa teaching their skills to the young Indian men. They and their families were all housed near the big agency store. We even had a reservation doctor and his family living two houses from ours.

The agency store gave out government food rations to all the Indians living on the vast prairie every Saturday. The Indians brought in animal furs and other Indian articles to trade for the much desired guns and whiskey. It was a very busy place.

On Saturdays the Indians rode out to a corral where steers were penned up. Cowboys got the steers running around in circles looking for a way out of the corral. Two of the cowboys stood in the center of the dirt area on a raised platform and took turns shooting them with a rifle. When all the steers were killed, the corral gates were opened and each Indian came in on his pony and dragged the steer he wanted to take home to butcher for his family. Occasionally a squaw would help. People came from all around the area to watch. Even newcomers on trips from the East would come.

Neither Mama, Nellie, nor Mabel would watch the steers be-

HATTIE

ing killed. They said it seemed cruel, but Fred and I watched every time; to us it seemed a natural way to handle the Indians' weekly meat allotment.

One day an old Indian wanted the men to let him inside the corral before they shot all the steers. He wanted to try his luck shooting down a steer with his bow and arrow. Everyone was anxious to see what kind of shot he was. Fred and I watched curiously. I thought at first that the old Indian would be trampled to death getting to the platform, but he swiftly and skillfully made his way and climbed up to where the cowboys were.

We all stood waiting while he took his time. The steers wildly raced around and around, stirring up clouds of dust as they looked for an escape route. The dust coated everything, making it hard to see. The old Indian was dressed in buckskin pants and a vest; it looked as if he had never taken them off. His black hair hung in a long dusty ponytail down his back. He stood motionless for a few minutes, and finally raised his bow. Hardly seeming to take aim, he let the arrow fly in one quick, swift motion. I saw a steer fall over. Everyone seemed to gasp as it fell. Several steers almost fell over it, which seemed to make the rest even more terrified as they raced faster. Soon the cowboys managed to kill the rest.

The crowd waited for the dust to settle. We were all eager to see where the arrow had struck. Finally the cowboys pulled the old man's steer out and no one could believe what they saw: the arrow had gone between two ribs and had struck that steer right in his heart. The old man puffed out his chest and raised his bow high in the air, letting out a whoop that could be heard above the noise of the crowd. The old Indian danced with pride.

That evening I asked Papa about it. "Why was the old Indian so proud, Papa?" "Well Hattie," he said quietly, "It's like this, Indians lived here long before our people came. They hunted, gathered food, and lived very freely. They had their own customs, religious beliefs, and even their own medicines. You know all that don't you?"

"Yes Papa, I know that, but why was he so proud of killing that one steer? I think that would be kind of easy for him being it was penned up, and at close range like that. It's not like killing a buffalo out on the plains."

Papa smiled and said, "That's good logic, Hattie, and you're probably right, but for the old Indian, killing his own meat with his own bow and arrow showed he still possessed great skills as a hunter. This gave him a feeling of pride once again. Can you understand that?" I looked at my Papa with respect, for he always seemed able to explain a mystery of human nature in a way I could understand.

Mama didn't seem to mind life on the agency. Of course it wasn't like living in Chamberlain, but she seemed to be happier and healthier.

"I am writing a special letter to Grandmother Taylor today." She announced to us one afternoon. "I have something special to tell my family, but I'd better tell you all first." She paused, looking at us with pride and a twinkle in her blue eyes.

She was sitting in her rocking chair in a worn blue cotton dress, her brown reddish hair combed tightly in a bun at the nape of her neck as usual. I loved to watch her comb her hair; it was thick and almost reached to her waist. It would have been longer, but she got Nellie or Mabel to cut it every once in awhile. Her eyes were blue like the sky on a bright sunny day. Today they were happy eyes, but they could get very cloudy.

"We are going to have another baby!" She announced to us proudly.

Nellie, Mabel, and I pretended to be surprised, but I wasn't and I knew my sisters weren't either. I had overheard them talking one afternoon while Mama was out hanging up clothes. I heard Nellie say to Mabel, "Did you notice Mama's stretching her back a lot, just like she did when she was expecting Elizabeth?

"She looks like she is putting on weight around her middle too," Mabel answered.

"Yes, and she seems kind of pale. I noticed she didn't eat breakfast this morning. Did you see her put her oatmeal back in the pot? She tried to hide what she was doing, but I saw."

They both moved away from where I was, so I didn't hear more,

HATTIE

but I began paying attention to Mama. I hadn't noticed anything different up till then. Soon I too saw that she seemed tired, lying down on her bed in the middle of the day.

So, Mama's news wasn't much of a surprise to us girls but it definitely was to Papa and Fred. Papa looked at Fred and grinned. "Maybe we'll get another boy." He went over and gave Mama a big hug. Papa loved children and always said, "The more the merrier."

He then turned to us children saying, "I want you all to double your efforts at work around here and help your Mama out." His voice was kind of stern and we knew he meant it and would be watching us. He, too, seemed to straighten himself out. He had slowed his drinking down a lot since we had moved, which had pleased Mama.

Explaining our new home in the letter to her mother, Mama told her that we had a group meeting for worship on Sunday. She knew this would please her.

Grandmother Taylor always showed up at church every Sunday with all her children dressed in clean, starched shirts and dresses, every one of them with neatly brushed hair and well-scrubbed faces and hands. "Yes Sir, Lord, we come to you clean outside and inside," she was fond of saying. Mama said that her mother inspected all her children before leaving for church, checking that the older ones had helped the younger ones prepare properly. Mama liked to remind us of that inspection, as she lined us up and did the same to us!

Grandmother Taylor had memorized many Bible verses and Mama could also recite them perfectly from memory when something would trigger a quote. If someone would say something hateful, Mama would quote:

> *"Ye have heard it said, Thou shalt love thy neighbor, and hate thine enemy,*
> *But I say unto you, Love your enemies, bless them that curse you, do good*
> *to them that hate you, and pray for them which despitefully use you, and persecute you. He maketh His sun to rise on the evil*

SHE WAS WIRED DIFFERENTLY

and on the good, and sendeth rain on the just and on the unjust.
  For if you love that which loves you, what reward have ye?"
(Matthew 5:43-48)

Mama had a faith of her own. She had figured out a way to live her life with a deep faith in God. Many times she told us, "Faith is quite different from hope or belief. Hopes can be dashed and even certain beliefs can be changed or proven wrong, but faith is eternal. Faith is stronger than either hope or certain beliefs; it is a deep inner feeling of God watching over us all." Then she would add, "Try your hardest to find faith. You will need it to get you through the struggles life brings."

As the weeks passed, it became apparent something was preoccupying Mama's mind. She seemed distracted and worried. One day Nellie was outside hanging up the wash and she overheard Mama expressing her concerns to the agency doctor. Nellie kept out of sight behind some bushes as she listened. She told Mabel, "I heard Mama telling the doctor about a concern. She told him she was almost forty years old and that the birth of Elizabeth had been harder than any of her other babies and that it had taken longer to recover. She told him she knew several older women who had died in childbirth and was worried about leaving us five children without a mother. She asked him for any suggestions about what she could do to safeguard her health while she was carrying her baby."

Nellie then said, "I was so mad because all that old doctor said to Mama was, '*most* women handle it okay, so why worry?' That was so insensitive, don't you think?" She almost sputtered her anger, "He just callously brushed away her fears! And he had no suggestions for our poor, dear worrying Mama!"

I became concerned after hearing this conversation and decided to watch Mama closely. I didn't have any idea of what to look for or what to do for her but I did begin to help her lift the heavy pots onto the stove. When I could, I would help water the garden.

I also started thinking again about being a woman and what it means. The more I understood about Mama's role in life, the more I questioned it. Did all women have to get married and bear one child after another in pain and suffering, while still being expected to cook and clean for long hours every day? I couldn't shake that thought. Why would anyone want to be a woman? Their lives seemed full of drudgery to me, while men's lives looked more and more appealing.

Papa had gotten us all ponies so we could all help gather firewood and ride the prairie for fun. Mama decided to make split skirts for us girls so we could ride astride in comfort and still be properly dressed and ladylike

The skirts allowed much more freedom of movement, and modesty, than the skirts we had been wearing, which had a tendency to ride up and show our lower limbs. I thought the skirts a vast improvement, but still felt they were a little too fussy for me. Mama promised she would make one for me after she finished Nellie and Mabel's, but she was often tired so put it off. I was glad, because I had a different idea brewing.

Mama made most of our clothes. Papa bought his and Fred's trousers from the general store. Fred had grown fast; he was the same size as me now. I was small for my age and Fred was big for his.

One day before we were going out on the prairie to find some firewood and gather some food, I said to Fred, "Bring the old trousers you outgrew, Fred, and keep it a secret from Mama." Papa had just bought him a new pair of trousers. His old ones had not been worn very much. He had outgrown them rather than wearing them out. Fred gave me a puzzled look but said "Sure." He always did what I wanted. We were buddies, but I was the leader.

When we were out of sight of our house, we stopped our ponies. Jumping down, I said to Fred, "Did you bring your old trousers?" He reached in his gathering bag and handed them to me, giving me a puzzled look. It quickly turned to wonderment as I slipped the trousers on under my skirt and began tucking the skirt into them. I had picked a plain top, so it looked like a shirt. I was satisfied I had smoothed the skirt as flat as possible into the pants, making me look a little bigger, which was good, I thought. From my gathering bag, I took an old soft leather cap I had found on the prairie a few weeks earlier. I shoved my hair up under the cap and smirked at Fred. "See, now I'm just another boy, your brother. I'm not your sister anymore," I laughed. "You may call me 'Cappie'," I said, strutting around for a few minutes. "No one will ever guess that I'm a girl, Fred." Jumping back on my pony, I felt the freedom of swinging my leg over its back and sitting astride as I had seen men do thousands of times. I thought, "Now this feels right! This is comfortable."

I began wearing Fred's old trousers every time we went riding, loving the freedom of movement this gave me. It was hard sneaking around. One day Mama caught me, just as Fred and I were leaving. With a horrified look on her face she exclaimed "Hattie, what are you wearing? You aren't going out in public like that."

I began pleading with her, begging her to let me wear the trousers while riding. "Mama, I am so comfortable. It's really just like Nellie and Mabel's split skirts. All their hand-me-downs are so worn out, they are threadbare in places. Please, Mama, I put my hair up in this cap and everyone thinks I'm a boy, Fred's brother. Isn't that safer for me?"

Mama stood with her mouth open, unable to speak. Papa had heard Mama's exclamation and came to see what was happening. I turned to him and pleaded, "Please, Papa, please tell Mama it's all right to let me wear Fred's old trousers. See, Papa, see how I look, just like a boy," I hurriedly stuck my hair up under the old cap to show Papa how I looked. "Fred and I have gone out like this several times now and everyone thinks we're brothers! This is so much safer than looking like a girl, don't you think so?"

Papa stared at me blankly for a moment. I was afraid of what he was going to say. But his face broke up into a huge grin and he just laughed and laughed. Mama looked more shocked at Papa than she had at me! When his laughter subsided, Papa smiled at me and said, "Well, my little intellect, you've found a way to be a boy! You certainly look like one right now, and I agree you are much safer if people you meet don't know you're a girl, and turning out to be a very pretty one, I've been noticing. Those old trousers fit you very well."

HATTIE

He turned to Mama and said, "Lizzie, I think we could let Hattie wear her brother's trousers. I believe she is safer when she does." I could tell Papa's words made sense to her, but it was still hard for her to think I would rather wear my brother's trousers than wear a split skirt.

I said, "I love to wear the trousers, Mama, and they are only worn a little, not like the hand-me-downs from my sisters." So, with both Papa's and Mama's blessing I began wearing Fred's hand-me-down trousers whenever I rode on the prairies, which was most of my time.

This was a momentous occasion in many ways, some of which I wouldn't come to understand until many years later. With permission to wear trousers and cap and pretend to be a boy, there seemed to be an unspoken permission to pursue my freedom. It was my first taste of how life would be if I had been born a boy.

However, I had to wear dresses when visitors came and when I went to school, to church, and posed for family pictures. The more I wore trousers, the more uncomfortable dresses became. Mama and Papa made me wear a dress for family pictures. Looking at me in those photos, anyone could see how uncomfortable I felt and how out of place I looked. I hated dressing up and I didn't like my older sisters fussing over me, trying to fix my hair. They would giggle together, telling me I must try to look attractive for the "young men." All I felt was strange and awkward, as though I was playing some make-believe game. It just didn't feel right, like I felt when I was riding out on my pony wearing Fred's trousers and my hair in the cap. I could not explain this to anyone, not even Papa. But I think he knew.

After living among the Indians for so long, Mama had become quite comfortable with them. She liked to talk with the Indian women and showed them how she made bread and cakes. In turn, they were quick to teach her about some of their medicines and their beautiful decorative artwork.

Mama had one special Indian friend who had more knowledge of herb medicines than the others. Her name was Sun. She made up a special Indian herb concoction for Mama and said it was given to pregnant Indian women to help them stay strong.

SHE WAS WIRED DIFFERENTLY

Sun helped Mama brew this tea, telling her to drink one cup twice a day until she delivered. After a week or two of drinking the tea, Mama did seem to feel better. Her face regained its color and she had more energy. In return for the herbal tea, Mama taught Sun how to preserve food for the winter the way she did.

One day Fred and I were heading back home from an unsuccessful morning of fishing. We saw Sun camped a short distance from our house. She had made a little tepee with a few tree branches and a large piece of Buffalo hide. I felt uneasy, wondering why she was camping there.

Nellie came running out to meet us looking more excited and concerned than I had ever seen her. I blurted out, "Why is Sun camped here? What's wrong? Is Mama all right?"

Breathlessly she said in an excited rush, "Mama is having the baby: Sun is here to help her. Mama wants you and Fred to help watch little Elizabeth. Take her for a walk down by the creek where we picnic. Mabel or I will bring some food out later. First take her for a walk to the agency store," she instructed us breathlessly. "Try to keep her from getting frightened. We don't want her to hear Mama crying out. I'll hang a white cloth outside the door so you can see when it's safe to come back."

"Where is Papa?" I asked Nellie after I had gotten down from my pony. Nellie had a frantic look on her usually calm face but it changed to one of irritation at my question. "I don't know," she answered in an agitated voice. "He didn't come home again last night." She usually didn't share her concern about Papa with me but I knew she and Mabel talked in whispers about him. At night in bed they would exchange angry glances on occasions when he came home. He had started drinking heavily again and often stumbled into bed. We knew recently he had taken to sleeping off his drinking bouts at work, where he had a cot to sleep on. Mama hated Papa's drinking. She would remind him of his Baptist upbringing, which was strictly against drinking. It had become a big problem between them again, and our whole family felt the strain. But until now neither Nellie nor Mabel had discussed it

HATTIE

with me. That Nellie was agitated enough to speak to me about her concerns showed just how badly shaken she was, which in turn worried me more.

We all had felt bad for Mama. We missed Papa as he used to be. I knew that most of the men around us drank a lot and gambled as well. Some of them hit their wives and children, gambled away all their money, and provided their families with little food. Our papa didn't get mean, so I always defended him in my heart. In fact, I told myself, he seemed to act very funny and happy then he would just fall asleep. Sometimes he fell asleep at the dinner table and we would help him to bed. We had learned we couldn't trust anything he said when he was drinking, and we couldn't depend on him either. What had started as a very occasional occurrence, usually after a big job was finished, was now happening several times a week. He had almost stopped after Mama announced another baby was on the way, but that had lasted only a month or two.

Nellie pulled herself together and spoke with authority again. "Don't worry about Mama. Mabel and I will help Sun take care of her and the new baby." Seeing the worried look on my face, she patted me on the shoulder and said, "Don't worry about Papa either; he can't help Mama now. I'm sure word has already spread about Mama being in labor. You know how fast rumors and news spread, Hattie, so Papa will stay away until the birthing is done; men don't like being around at times like these."

Listening to Nellie, I realized I had been elevated to a young woman who was now included in women's activities. I looked down and saw Elizabeth looking a little frightened. I reached down and took one of her four-year-old hands. Fred took the other, and we put smiles on our faces for her. We could tell she was puzzled, wondering what was going on. That was when I heard Mama cry out, so I began to walk away, as quickly as little Elizabeth's legs would go. I looked back for just a moment, long enough to see a frightened look come over Nellie's face as she turned and hurried back toward the house.

Walking toward the store, which was about a half mile away, we first

SHE WAS WIRED DIFFERENTLY

passed the little vegetable garden that Mama planted after we moved to the agency. She planted a vegetable garden at every new location we lived, if she could. She loved vegetables. I remembered watching her bending over the rows, weeding and watering them with tender care. She lugged a big watering can along the rows. We saved all our bath and dishwater from the house to water the vegetables. I learned at a very young age that nothing was wasted.

"Why must women's lives have to be so hard?" I wondered. It didn't seem fair to have to endure this terrible childbirth too! How unfair of God, I thought. Then in the next moment I begged for forgiveness, and asked God to take care of my Mama.

I could still hear those first cries of anguish ringing in my ears as I hurried Elizabeth and Fred on even faster. I kept trying to reassure Fred that Mama was going to be all right, as much to reassure myself as Fred. I wasn't sure about it at all. What would we do without Mama, I wondered. This gave my heart such a twisted, sick feeling, I thought I might throw up. So I made myself think of our job to help Elizabeth so she would not get scared.

When we reached the agency store, Fred went to look at a small rifle he had been trying to convince Papa to buy him. I walked Elizabeth around the store, looking at all the displays and trying to appear nonchalant. Mr. Hallman, who ran the store with his wife, finally asked, "Anything special I can get you kids?" I replied, "No, sir, we are just looking and showing Elizabeth around the store."

Mr. Hallman's wife had heard about Mama, however, and she smiled at us. "I was just telling Mr. Hallman some of our stick candy is getting stale. Perhaps each of you would like to have a piece?" Candy was a big treat for us, one we only were able to indulge in two or three times a year on holidays or birthdays. Fred and Elizabeth quickly replied, "Yes, please!" They stuffed the candy stick in their mouths as soon as she handed it to them. Somehow I just did not feel like candy and decided to save it, so I put mine in my pocket. The day before, I would have been sucking on the candy just as greedily as Fred and Elizabeth, but something had happened to me that day. I think I had taken a big step toward adulthood at thirteen.

When we left the store, I hesitated for a moment, wondering where we should go. I couldn't bear to go by the carpentry shop, couldn't bear to see Papa drunk or sleeping off his drink. I headed for a pretty, shady spot down by the creek where our family liked to picnic. I sat with my feet in the creek stream while Fred and Elizabeth finished their candy, then I carefully washed Elizabeth's sticky mouth and fingers in the water, as Fred splashed around. Soon Elizabeth started getting fidgety, so I suggested we pick some pretty flowers for Mama. The bouquet would stay fresh if we put the stems in the water, I told her. Fred found an old tin can and filled it with water, helping me keep Elizabeth entertained. Although he didn't say so, I could tell Fred was worried about Mama. He was being very quiet and helpful with Elizabeth.

Late in the afternoon we walked slowly back toward the house, hoping there would be a white cloth on the door. But no white cloth was there. Elizabeth was very tired and fussy, so I sat down with her on the big wide swing Papa had built us. She loved to swing and quieted down as soon as Fred started pushing the swing from behind. He started pushing harder and harder until I had to speak to him sharply, telling him to stop. Then I pumped the swing slowly with my feet while Elizabeth fell asleep with her head in my lap. I stroked her hair slowly, thinking what a sweet little girl she was. Fred went to play in the dirt near the garden, drawing figures with a stick. I closed my eyes, just for a minute, I thought, but the next thing I knew Nellie was shaking my shoulder saying, "Hattie, Hattie, wake up, we have another baby sister!" I stood up, shivering from the cold, saw it was dark and the moon was already out overhead. Nellie carefully picked up Elizabeth without waking her.

"How is Mama?" I asked hurriedly as my concerns came flooding back.

"She's doing fine, the birth went really well. Mama's a real trooper!" Nellie proudly said.

Inside the house I saw Sun holding a very small bundle, a big smile on her toothless face. "Hey, see girl! Baby and Mama both good! Look,

look!" She bent down to hold the bundle at our eye level and uncovered a small bright red face that looked like it was preparing to cry. The little face was so scrunched up and comical we all laughed. "See, baby crying, hungry!" Sun said proudly, as if she had given her birth and was the new mother herself.

I couldn't believe anything that small and helpless could live. I had forgotten how newborn babies looked. I thought she was very ugly, like a freshly plucked chicken. Was this what Mama had gone through so much pain and suffering for, this tiny little thing? Did I look like this when I was born, I wondered. I hoped Mama thought little Ethel, as she was to be named, was worth it.

We girls all told Fred we were sorry it wasn't another boy for him to play with, but he didn't seem concerned. "A baby boy would have been so much younger than me," he said, "that I really couldn't play with him anyway. Besides," he laughed, "I have Hattie, I don't need another brother." I grinned back at him. He enjoyed playing "brothers" with me and didn't care that I was really a girl.

Soon we were told Mama wanted to see us so we all trooped into the bedroom. Mama looked a little pale and weak, but she gave us a big smile and held her arms out to give us all a hug and kiss. Elizabeth remembered the flowers in the tin can. She ran and got it off the porch, and ran back to present it to Mama with a big smile. Mama gave her an extra hug and said, "They are very pretty, Elizabeth, just like you!" This got her a big smile. Soon Sun came in to shoo us out, saying, "You kids, get out now, Mama rest!" Mama smiled and said, "Sun is right, my dear children, I need to rest a little now. In the morning I will feel better."

As tired as we all were, Nellie and Mabel made us a late supper. Then Fred and I both fell into our beds. Nellie lay down on top of her bed still fully clothed and was instantly asleep. Mabel managed to get into her nightclothes before falling asleep.

In the morning Papa came home anxious to see his "beautiful Lizzie" and his new daughter Ethel. He looked tired; his clothes were all wrinkled because he had slept in them. His beard was unkempt and

HATTIE

so was his hair. Papa seemed worn out too. I didn't know what had happened to the Papa I remembered as a little girl. I wished for that Papa back. If only I could do something to help, to change Papa back the way he was and make Mama happy again. Prayer seemed to be my only option, the best one as Father Willard had told us so many times. "When you have a problem, turn it over to Jesus." So I prayed.

CHAPTER FIVE

We spent several years on the agency. Eventually Nellie and Mabel learned all that the teachers at the agency school could teach them. We still had our house in Chamberlain, so Papa and Mama decided that Nellie and Mabel would go to school there during the week. A boatman ferried them across the river between the reservation and Chamberlain for the weekends.

When the girls went to Chamberlain, Mama was left with more work than she was used to. I tried to help the way my sisters had, but I was clumsy, burning food and breaking things. It seemed I just made extra work instead of helping. Mama finally threw up her hands and said with a sad smile, "Hattie, it's obvious cooking and housework aren't for you. You try hard. You are good at things like bringing in firewood and water, getting the fire in the stove started, finding rabbits and fishing the creeks out on the prairie. Somehow that suits you. I guess you'd best stick to what you are good at. Like Papa has always said, you are very intelligent; still, I worry about your future. I think you are going to live a different life than your sisters."

Mama's words caused me both pain and pleasure pain, in being judged "different." Pleasure, in knowing that Mama understood me. I knew Papa did. I thought perhaps I could make up for my lack of ability, so I blurted out quickly, "Mama, you are right. I know a way I can

HATTIE

still help you. Besides the chores I have been doing, I will take care of the vegetable garden and Fred can help me." Fred nodded as he listened to our conversation. He was always at my elbow, trying to help me. I explained, "I've watched how carefully you weed and water between the rows. I'm sure I can take care of the garden properly. That will save you a lot of time and work, won't it?" Mama thought a moment. Looking at me with a serious face, said, "Hattie that certainly would be a big help to me, but that garden is very important to our family. You must keep the watering up and notice when bugs are taking over. And there is the job of sprouting new seeds and knowing when to plant. It might look easy, my dear, but there is lots to do." She was looking at us with doubt, and continued explaining, "The family needs the food we get from the garden. You both understand that don't you?"

"Yes, Mama," I said. "We will not let you down; watch us, you'll see."

Fred piped up, "And we'll be learning about plants and the seasons like you know them, Mama"

She gave us each a big hug, and said, "All right I will trust you two, and I'm going to tell your Papa about this. I'm sure he'll be proud that you are helping out."

With Fred's help, the gardening only took a few hours a day; we still had plenty of time to ride our ponies. Spending so much time working and playing together built a close bond between us, more than just brother and sister. I didn't know how much I would cherish the memories of those days roaming the prairies, exploring all the nooks and crannies. As soon as we left the house, I was a boy enjoying the same freedom as my brother. We looked alike and invented the name Alfred, which I thought suited me well. It was Papa's middle name.

Fred and I rode into the hills and watched herds of buffalo, fewer in number each year, as they thundered across the prairie, stopping to water when they reached the Missouri River. Other times, we would race across the prairies ourselves, scaring up prairie chickens and anything else in our path. Sometimes we would see a sod shanty and tepees off in the distance and would ride over to visit the Indians. Papa had taught us to greet the Indians with "How Cola" which he said meant, "How do you do, friend." "Follow this with a handshake. That way the Indians know you want to be friends," he advised us.

The way Indians lived was fascinating and very different from our way. We watched squaws make moccasins, vests, and papoose bags for carrying their babies on their backs. Nothing was considered complete until it was decorated with colored beads, generally made from porcupine quills. They were fashioned by the squaw biting down on a quill two or three times, sliding it flat between her front teeth. After the quill was shaped satisfactorily, squaws colored them with one of several dyes made from plants. They then cut the quill into beads. We were told that each tribe had its own secret colors from the plants they used. Exactly how the bright colors were made were tribal secrets.

Papa taught carpentry to the Indian men six days a week. He started very early in the morning, came home for lunch, and then returned to the carpentry shop until suppertime. After Ethel was born, he didn't go out drinking very much anymore. It seems he wanted to please Mama. One day Papa came home for lunch looking very sad. He stood looking at Ethel in her baby bed for a few minutes, then said to Mama, "One of my students, a young man, lost his baby son last night, Liz. I feel so sorry for him. We are lucky that Ethel is a healthy baby. I've spent the morning making a small coffin for his little one; I felt I should do something for him. He has been helping me with it, and I think it is making him feel a little better."

Mama sat for a moment, looking very sad. Her expression suddenly changed and she stood up and said with determination, "Well, Melroy, I think the coffin is a wonderful idea. It just needs to be finished for that poor little baby."

Papa looked at her quizzically, "I'm not sure what you mean, Liz. We have sanded and stained the coffin. What else can we do?"

Mama said, "I've got some nice soft cotton, perfect for lining the inside." She hesitated for just a moment before she hurried to her big cedar chest where she kept all her prized possessions. She pulled out a bundle of cotton and also her white satin wedding dress. "I've been sav-

HATTIE

ing this for one of our daughters to wear one day", she said, "But both Nellie and Mabel are much taller than I am. It would never fit them and I know Hattie is too big for it now. This is a special occasion, so I'm going to fix a piece of the satin into a pillow for the little one's head to lie on." Papa looked unsure, but Mama said, "I think the Good Lord wants me to do this. It will help the Indians understand our Christian ways of sending our loved ones on, and besides, if it were one of our children, I would appreciate the kindness of those who wanted to share our heartache."

I watched as Mama spent the rest of the day and evening taking out the stitching in her wedding dress, pressing the heavy satin material and making a small pillow. Mama spent the afternoon at the shop with Papa, padding and lining the coffin. When they were satisfied with their handiwork, Papa took the coffin to the grieving parents. When he returned, he said, "Everyone inspected the coffin inside and out. They all felt the pillow and were so pleased; I don't think they had seen fabric so soft and beautiful. They thanked me over and over. They said: 'Good, good' and patted me on the back. Each one seemed to want to touch my hands."

From then on, the Indians came to Papa when someone died wanting him to make a coffin. Papa couldn't bear to turn any of them down. Mama ran out of the white satin and had to look for alternatives. I had gained a deeper respect for Mama, giving her cherished wedding dress out of love for a little Indian baby that she did not even know. Many people still looked at the Indians as savages, but Mama and Papa saw them as children of God. Fred and I had come to respect the Indians too. We knew Papa and Mama were very smart in many ways and were good and decent people, but seeing them do these kind and loving things for the Indians made me even more proud of them.

Papa decided to open a print shop on the agency and teach the young Indian men how to print newspapers. The Yankton Indians had printed a Sioux-language paper, called The Lapi Oaye, at Greenwood on the Yankton Reservation not far from our agency reservation. Papa had helped with setting the type in the Sioux language.

SHE WAS WIRED DIFFERENTLY

Fred and I were quick to volunteer our services. We wanted to learn the newspaper business from top to bottom, and this was our chance. Both of us discovered we had more than an interest; we also had a natural aptitude. We treated each edition as a big puzzle, fitting everything neatly into place on the page. Papa patted us on the back and said he was proud of us.

Cleveland took over the presidency and recalled all the quick-claim lands. He said the Indians should have first choice of land because it was their homeland. This caused lots of trouble. And now, several years later, he had to send the army in to enforce his orders.

The army built some barracks right down on the river's edge near the agency. The settlers grouped together to fight the army, but soon the army enforced the president's orders and things finally settled down.

While the army was stationed in their barracks on the river, our house became a gathering place for many of the young soldiers. Nellie and Mabel enjoyed the attention of these men. Mabel played the piano. She had been taking piano lessons, and Nellie took painting lessons from two of the reservation wives. This was our first social gathering of singing and dancing with young men under the watchful eyes of Mama and Papa. Fred enjoyed dressing up in bits of uniform that the soldiers gave him; Papa even took a photograph of him dressed up.

I was eager to talk to the young men and ask them about military life, but I soon gave up. They seemed to enjoy my older sisters' company the most and would kid around with Fred and treat him like a young brother. They told him stories about their lives and he listened eagerly. Again I envied Fred. He was accepted, and even though I dressed like a girl, the soldiers paid no attention to me. I felt awkward and out of place.

The memories of our last year at the agency, I never forgot. We had lived so long among the Indians, we felt safe among them. Then one day word came that the Sioux Indians from the Black Hills Rose Bud reservation, just across the Missouri River from us, were on the warpath.

Our feelings of peace and safely were shattered and replaced by fear and distrust. Many of the white people on the agency left. Papa said he wasn't worried about us at first because we had so many good friends among the Lower Brule Lakota Tribe. But the Sioux in the Black Hills were from a different group of Indians than the Sioux we lived among.

As the resistance continued day after day, Papa said, "We can never be sure how safe we are. After all, we have taken their land, forcing them to live on these restricted reservations when they were accustomed to roaming free. Many of the huge buffalo herds have been slaughtered, and buffalo was what the Indians had depended on for food. Our government has made many promises, only to turn around and break those promises, and now Sitting Bull has been killed. We would be outraged if the roles were reversed."

Papa often spoke like this to anyone who would listen. It wasn't popular with the settlers, but he didn't seem to care.

The Rose Bud Sioux and the soldiers had several skirmishes in the Black Hills and rumors among the settlers flew. One rumor that frightened us was that after a fight, the squaws would run out and scalp the dead soldiers. Another was that our people found Indian babies, still alive, next to their dead mothers. All were rumors, of course; we didn't know which were true and which were not. The young soldiers reported finding bundles of scalps on some of the Indians, and that scared us plenty.

Night after night we were kept awake by the sounds of many horses and war whoops from across the river. Mama feared the war cry of the Indians. "It goes right through a person like a cold chill, freezing the blood in your veins," she said, with a shiver. "I can hardly stand it, and I will never forget it." We all knew what she meant!

Papa arranged for the boatman to be ready at a moment's notice to ferry our family back to Chamberlain if the fighting across the river would somehow set our Indians off. He sent word to Nellie and Mabel to stay in Chamberlain. They sent word back that they could barely sleep wondering what was happening to us. Eventually the soldiers were able to put down the unrest on the Rosebud.

Soon after the uprising, Papa and Mama decided to move back

SHE WAS WIRED DIFFERENTLY

to Chamberlain. Nellie and Mabel were thrilled. Fred and I were not very happy at the change since it meant we would no longer be able to wander the prairies on our ponies. I would have to give up wearing pants and wear dresses every day. This thought made me unhappy. I knew I would have to go to school with people I didn't know. In some ways I felt superior to them. I had learned so much about nature from the Indians. In other ways, I felt inferior, and I knew I was especially shy around children my age; it was painful to be judged as different by others.

Another family portrait was taken about this time 1890, and it showed how much our family had changed. Nellie was eighteen, Mabel sixteen, Fred eleven, Elizabeth eight, and baby Ethel was four. Comparing Mama and Papa to the picture taken ten years earlier and another five years after that was very revealing and told of the hard life we had lived. We had aged a lot in the last five short years. We all looked tired and sad. I looked angry. I asked myself why. I had no answer to that question, but I knew that the camera didn't lie. I was angry, and didn't know why.

Papa began doing carpentry work again. He got very busy with the many new people moving into town wanting homes and businesses built. He hired a man named Emory Knotts to assist him. Emory and my sister Mabel fell in love and asked Papa's permission to marry. Papa insisted they wait until Mabel turned eighteen. She was only sixteen. Both agreed to the two-year engagement. Mabel began to get household goods together to start her own home. Papa made a cedar chest for her. She began filling it for her marriage. She and Nellie were delighted and swooned over each pillowcase and tea towel. I thought it was all very silly.

One day Emory was reading the Chamberlain Chief, Papa's old newspaper, and saw an ad offering a newspaper for sale in Salem, South Dakota. By this time our area had gone from the Dakota Territories to statehood as South Dakota. Papa and Emory had talked for some time about wanting a newspaper. Papa was tired of carpentry and said that he would love to be back in the newspaper business. Emory was all ears. He wanted to learn a trade, he told Papa, because he didn't like the building business either.

"If we find a newspaper for sale, Emory, I'll teach you the business," Papa told him, so when Emory showed Papa the ad about a newspaper for sale in Salem, they got very excited. They looked up Salem on a map and discovered it was a tiny speck of a town, a little south of Sioux Falls, one of the biggest cities in South Dakota. Emory gave Papa a mischievous smile and said, "Let's toss a quarter to decide: heads we buy the paper and tails we stay here." That's how our future was to be decided, with the toss of that coin! We all held our breaths wondering what course our lives were going to take.

The coin came up heads. I glanced quickly at Mama to see how she was taking this news. She seemed happy with it. I think Nellie and Mabel were the happiest. Mabel could see that Emory had ambition. She knew he did not like the carpentry trade and she too liked the idea of the newspaper business as their life's work.

Papa and Emory traveled to Salem to look over the newspaper office and the town. They decided that both would suit our family nicely. Emory had saved three hundred dollars, and Papa put up the rest for the purchase, forming a partnership. They sent a message back to Mama telling her the news and saying "Come, Quickly!"

Mama was glad to leave Chamberlain, which surprised me. I mentioned it to Nellie, who whispered to me, "Papa's drinking is well known around town now and some of Mama's friends have seen him coming out of the saloon drunk. Most of Mama's old friends have been avoiding her because of Papa's drinking. It hurts her feelings." I thought, after all the years of isolation at the agency, poor Mama had been looking forward to Chamberlain and the civilized life she had always wanted. Instead she sat at home with no callers and no invitations. Even at church, the other women pretended they didn't see her and whispered behind her back. I was stunned. It hadn't occurred to me that other people might judge Papa so harshly. No wonder Mama was happy to leave. She probably hoped Salem would be a new start for our whole family

SHE WAS WIRED DIFFERENTLY

When we arrived in Salem, we found that Emory and Papa had purchased a large old building three blocks from the newspaper office. Two girls had been operating it as a boarding house. The girls were tired of the business and looking for someone to take it over. There were five men renting rooms and boarding there. When Mama discovered that she was expected to operate the boarding house, she was shocked. Now, there were even more mouths to feed, more rooms to clean, and more laundry to wash. I saw a flash of anger and rebellion cross her face. She, who never raised her voice or spoke in anger, still kept her silence. After a few moments, she quietly started directing us where to put the boxes and barrels we were unloading from the wagon.

When everything was arranged to Mama's satisfaction, she turned to Nellie and Mabel. "Girls, there is no way I can possibly take care of so many people by myself. I will need your help every day. You will have to do the weekly laundry, keep the rooms clean, and wash dishes and the clothes. I can handle the cooking, at least most of it. Elizabeth's old enough to watch Ethel for me, which will be a big help. Papa said wood for the stoves would be delivered when we need it, and there is a store in town where we can buy food." Papa said he could use Fred and me at the newspaper office.

Fred and I were thrilled! For the next few years, Fred, Elizabeth, Ethel and I attended school in Salem. After school Fred and I headed off to the newspaper. What a wonderful time! Papa taught us along with Emory how to run a newspaper. We learned to typeset. Mabel also helped at the paper when Mama could spare her, which was often because Mama knew Mabel wanted to learn the business too. We learned how to load the paper, ink the press, and print the papers. We delivered them too. Papa beamed when one of us came running with a news item, something we had heard around town with our youthful ears.

I came to realize how much I wanted to be a reporter who went everywhere and talked to everyone. How much more exciting than Mama's life, I thought. If I learn everything about the newspaper business maybe I can get a job working on a newspaper in a big city. A paying job! I could be independent! These thoughts made my heart jump with joy. This then became my dream and my goal. I talked about it to Fred as he also was thinking about working in the business when he graduated from high school. I began pushing myself to learn to socialize with other people because I knew a reporter had to mingle. I tried hard to develop this skill because I knew a lack of social grace was a weakness of mine.

One day Fred laughed and said, "Well, Hattie, if you can't get a job with a newspaper because you are a woman, you could dress up as my brother again and work as a man." We laughed, but the germ of an idea had sprouted.

We had been in Salem almost two years when Mabel turned eighteen. She and Emory Knotts were married on her birthday, June 15, 1892. During this time Nellie was being courted by one of the boarders, Daniel Carey, who worked for the Great Northwestern Railroad. On July 27, 1892, a few weeks after Mabel was married, Nellie and Dan were married. This, of course, left Mama with no help in the boarding house. So she closed it down, and we moved into the Diamond Hotel.

CHAPTER SIX

Papa and Emory ran The Salem Pioneer Register. Papa began teaching Emory about the business. Mabel knew a lot about the paper business but Emory wouldn't take any direction from her; he wanted Papa to teach him. Fred and I would run into the office breathless from school; "Hi Papa, Hi Emory, can we set type today?" we would ask. "No" Emory answered, before Papa could reply. We could tell Emory was annoyed by our interruption, but Papa always smiled at us. He would dig into his pocket for a few coins, give them to us, and say, "Go for ice cream kids, come back a little later and I'll have an important job for the both of you." That was Papa! He would pat Fred on the back and give me a tight hug with his big strong carpenter arms. "Off you two go."

Our parents seemed happier in Salem. Nellie and Mabel were expecting babies shortly after they were married. Nellie had a nice little home, in the heart of Salem; she seemed very happy. Dan Carey worked as a section foreman for the Great Northwestern Railroad. He had invested money in land and in some buildings in Salem. Nellie and Dan were well-fixed financially. She had planted a garden, just like Mama did, and was raising chickens. Her plan, she told Mama and me when we visited her, was to sell the chicks, as well as the extra eggs, giggling mischievously. She said she was saving up to buy a piano for Christmas, and take lessons.

HATTIE

Of all my sisters, I admired Nellie the most. She was like Mama except she had more backbone; she knew what she wanted and went after it. I figured nothing was going to stand in the way of her getting what she wanted and thought "Good for her!" Dan was in love and in awe of her; he just stepped aside and followed her lead. Whatever Nellie wanted was all right with Dan. He had been Catholic all his life but when they got married, Nellie would not join the Catholic Church, so Dan gave up his religion and joined her as a Methodist.

I didn't think Mabel had made a very good match for herself. I could tell that Emory was crazy about her, but Fred and I did not like him. He was moody. Sometimes he seemed like a great guy, but other times his mood was mean. I didn't trust him. I could tell Mabel seemed afraid of his moods. She was quiet when he was around. I discovered he was drinking; he tried to hide a small flask in his boot. I would see him pull it out, when Papa wasn't around, and drink from it. I knew he didn't like me because I couldn't or wouldn't hide my dislike, distrust, and mostly my disgust of him. He seemed always to be watching me out of the corner of his eye. Sometimes he would get very close and put his arm around my waist when Papa was not there. When I would move away and brush his arm or hand away, he acted hurt. He said one day, "You don't like me much, do you Hattie?" I said, "No, Emory, I don't." Then I moved away from him again just as Fred came into the newspaper office. I could tell by Emory's eyes that he didn't like my answer, but from that day on I avoided him and he avoided me.

We were not disappointed the days we came into the office and Emory was not there. They became more and more frequent as the months sped by. Fred and I learned that he was looking for a newspaper to buy in another town.

The days we came when Emory was not there, Papa would let us set up a section of the paper by ourselves, or check a section that Emory had done. If I found a misspelled word or incorrect grammar, Papa would praise me and say, "Well done my little Chickadee" and do a little shuffle dance around me.

We loved our Papa very much, especially when he was in one of his silly moods. He hadn't been drinking much since we had moved to Salem. He kept busy; besides the weekly newspaper he was building a church.

When I would set type, Fred would be at my back watching my every move. He was almost as good as I was and he practiced whenever he could. The best time for us was after the paper was printed for the week; then we could put the type set away. We made a game of seeing who could do each section the fastest.

"Papa," I said one day, "I do so love to set up the type. Do you think I'm good enough to get a job when I graduate?" "Shucks honey, you shouldn't even have to ask me. You are as good, or better," he whispered, "than Emory!" I was delighted with his answer because I too thought I was just as good and maybe a bit better. I knew I was not yet as good as Papa, and maybe I would never be as good as him. He seemed to know everything. But I knew I was a much better speller than Emory. He knew it too because he would have to ask me how to spell words. I knew this bothered him, but he had to ask me or use the dictionary. This slowed him down and sometimes he still had to ask me something. I was secretly tickled.

My hope and dream was to find a job on a big newspaper like the Chicago Tribune or the New York Times. One day Fred told my dream to Emory, and he scoffed. "Fred, she is dreaming; they don't hire women on those big newspapers," he said. Fred told me what Emory had said. I asked Papa if that were true and he said, "Have you ever heard of Nellie Bly?" When I answered no, he said, "She is a newspaper reporter in New York. Look up her story in the library and write a story about her, and I'll publish it in our paper with your name as the writer."

I couldn't wait to get to the library to do this research and write about this remarkable woman. At first I wanted to show Emory, but as I began reading and writing her story I began to appreciate what Nellie Bly had done. She gave me hope for women.

I found Nellie Bly's real name was Elizabeth Jane Cochrane Seaman, and she was born in 1864. She wrote for the newspaper *World* in New York City. She had made a name for herself by being sent to the Blackwell's Island Insane Asylum, where she had herself committed insane. She was to write an unvarnished narrative of the treatment of the patients and the methods of management in 1887, about twenty years ago.

I wrote the story about this remarkable woman writer and Papa printed it with my name as the author. Papa knew how to motivate us. Reading about this newspaper woman had given me hope. I discovered that Nellie Bly was very unusual because women were being discriminated against in every line of work, and they were not allowed to vote.

Some women were able to become doctors and some were being put in high positions, but few were getting a fair share of the American dream.

This, then, began my crusade and my interest in women's rights. I found a group named Women Suffrage that was fighting for women's causes: the right to vote, the right to work, and the right to get equal pay. They had a long list of grievances. I began talking to Nellie about these issues. She seemed interested in women's rights too but mainly she was against drinking.

I began to look for any news about Susan B. Anthony and her fight for women's rights. Nellie was interested in the Women's Christian Temperance Society, called the WCTU; this group of women was against drinking. They wanted to outlaw liquor. That would require a constitutional amendment, Papa said, and he thought it would never get passed. These women were determined to do just that, amend the constitution. Their number grew, and even men joined this cause.

I remembered reading the history of Susan B. Anthony. I laughed when I read about her strong-willed behavior and her fight for equal rights. In her friendly town of Rochester, Susan B. Anthony with her friends Guelma McLean and Hannah Mosher and a dozen other women registered for the election of 1871. "Six years before I was born," Hattie thought, On Election Day, fifteen of these women succeeded in voting. Susan B. was arrested and found guilty of voting. She was ordered to pay a fine of one hundred dollars and replied, "May it please your honor, I will never pay a dollar of your unjust penalty." The fine remains unpaid to this very day. I laughed when I reading this.

Mama said she had pluck, and admired her too. Mama, Nellie and I had many long discussions about women's rights. I complained to everyone who would listen of the injustice women endure by not having the right to vote. I understood how the Negro felt; the civil war had been over for over thirty years and still the Negro did not have freedom. I felt the injustice discrimination they were also quietly suffering. Mama and Nellie's heart was in getting rid of alcohol and the evils of the drink, but I was more interested in getting justice and the vote for women.

This interest in the suffrage movement became my religion. In my diary, I wrote in detail of my passion for the right to vote. In page after page, I wrote of how ridiculous the arguments of the anti-suffragists were. Their arguments were based on their conception of the nature of women; they assumed we were a different species. I wrote about this because I never wanted to forget how very degrading their arguments were. I became outraged when I read statements in a handbill being passed out in—of all places—church.

I wrote the entire handbill into my diary, so I would never forget my rage. I never wanted to forget how this made me feel: like a second-class citizen. I showed the handbill to Papa. He read it and laughed, saying, "Anyone with any sense does not believe this nonsense, Hattie." I became even more frustrated. Doesn't **anyone** see this injustice and what it does to women's self-esteem?

Seared into my memory was one of the arguments of the anti-suffragists: "A woman's emotional instability would make her a dangerous voter. She would let her feelings rather than her intellectual concerns be her primary reason for voting," and, "Since women obviously could not be trusted to behave rationally, they would be extremely dangerous in a political setting."

The arguments the antis stressed against women voting stated that women were intellectually inferior and could not make educated decisions. The final despairing remark: "Women did not have the intellectual capacity of men because their brains were smaller and more delicate."

I was appalled by these arguments and wondered why so few women were not speaking out at the way men were portraying women. It was as though we were just a little higher in intelligence than animals and that we needed a male to think for us.

This became my passion, to somehow educate people. I had watched Mama work as hard as any man and I had watched the Indian women work as hard as the Indian men; in fact in my opinion most men would be at a loss without women.

This kind of thinking occupied most of my time. What could I do? How could I make a difference? This became my obsession.

I wrote a long article about these arguments made by some well-known anti- suffragists and showed the paper to Papa. I said, "Please read these ridiculous statements about women, Papa, and they are being made by some very well-educated men. Let's print a column in our paper about how stupid these statements are." Papa read my article, handed it back to me, and said, "Honey, I would love to print that exactly as you wrote it, but I can't; it would be stepping on too many toes and we want to sell papers and stay in business, don't we?"

This was really the first time I saw that we had to "play politics," as Papa called it, and I felt a bit disappointed in my Papa. It was again learning to keep my mouth shut and bite my tongue, as Mama had told us to do many times while we were growing up.

Fred began saying he didn't know if he wanted to go into the newspaper business. He loved the outdoors and especially horses. He never forgot how free he felt when we were riding the prairie on our ponies. We talked every day on our walk from school to the newspaper office. He said, "I don't like being confined inside a building Hattie, even in a newspaper building." I thought how lucky he was to have a choice at what he wanted.

Summers were very hot in Salem. It was nothing to have the temperature rise above 110 degrees. Fred and I walked slowly to the newspaper office on those hot days. We would stop when we passed someone's house where we could get a cool drink from their well. We would lower the bucket down into the deep dark hole. Pulling the rope up slowly, taking the tin cup tied to the bucket, we would scoop up a full cup of the cold, sweet well water. We'd take a long drink and follow it with a full cup poured over our hot heads.

Sometimes we sat on the benches that surrounded the well. We talked about our plans. I remember the many days Fred would say, "Hattie don't you wish we were back out on the prairie on our ponies? Don't you miss it like I do?" "Yes, I do miss it, little brother." That's what I still called him even though he was much bigger than me now. "But I am glad Papa is teaching us the newspaper business because we could not make a living out on the Indian reservation riding our ponies. We have to think about how to make a living, Fred."

Fred felt sorry for me. He realized my dilemma. He knew my dream about working for a big newspaper. He was now fourteen and I was eighteen, but we both seemed old for our age. One afternoon I remember I said, "Little brother, you will not have a very difficult time getting a job in the newspaper business, but you know I will. I don't want to marry a man just to have someone to take care of me and yet this is about the only choice I have."

Fred looked at me with concern. "Hattie, as long as I'm alive I'll take care of you – we could get a newspaper together, how about that!" he said excitedly, as he thought about this. I reached over and tossed his wet hair with my hand and said, "You'll fall in love with a woman who won't want me around." I laughed, choking back the tears of love I had for him.

I turned nineteen, and should have graduated, but I was still lacking in some of the subjects. I was ahead in English, but I was still behind in math. Papa and Mama talked it over with me. They encouraged me to go one more year. Fred was now fifteen, and they felt if I would stay he would stay too. He had been talking about leaving school. He wanted to find a job somewhere. He felt he would be one less mouth for Papa and Mama to feed.

We had moved to a boarding house where Mama was helping out

HATTIE

in the kitchen. Soon I was helping almost full-time in the newspaper office as Emory and Mabel had found a paper to buy in Spencer, South Dakota.

Emory had been having a hard time letting Papa run the paper. He wanted more say-so. Papa and Emory parted as friends, but were never the friends they had been in Chamberlain. Emory had shown that mean streak, especially when he was crossed on some issue, and we weren't sure how he was treating Mabel. He had not drank in Chamberlain; in fact, had helped keep Papa sober when they first met. But now he was the one with a drinking problem, it seemed.

I heard Mama talking to Nellie about poor Mabel. She said, "Emory has taken up drinking." I heard Nellie say, "The one thing, Mama, I would not put up with is drinking!" I saw the smile on Mama's face and knew what she was thinking, "Women put up with drinking and many other qualities in men because they have too," I thought, reading her mind.

Fred and I, along with Papa, were glad that Emory was leaving. We would miss Mabel and their new daughter and we knew Mabel was expecting her second child, but we wouldn't be missing Emory. Papa really needed Fred and me now. I decided to stay in school for the next year and help Papa with the paper. I was glad for the opportunity; I knew I would now get to do much of the work that I could not do while Emory was there.

I learned the newspaper craft in those few years in Salem. One part of the business which surprised me was how a woman would be treated so differently from men. I found this out by the way customers treated me when they came into our office to place an ad.

None seemed to think that I could help with their ad. They always asked for Papa. This I could understand, but I was surprised that as Fred got older and bigger, they would allow him to decide what and how to say something in their ad, and much to my surprise, would give him their OK to print. I never got this kind of respect from a customer. The most surprising aspect of this attitude, I found, was that it was women as well as men who showed this lack of confidence in

SHE WAS WIRED DIFFERENTLY

me. The customer would always say, "Now have your Papa check the ad before you print it, won't you Hattie?" They seemed to treat me as a less skilled person only because I was a woman, as far as I could figure.

One day, Mabel came to visit Mama, and I cornered her, because I knew that she worked in their Spencer newspaper office with her husband, and had even been to linotype school. I asked her, "Do the customers let you take their ad orders and give you the OK to print the ads?" She looked at me with a knowing smile, "No, Hattie, they just don't trust 'us' to say, print it!" She saw the disappointment on my face, and putting her arm around my shoulder, said, "Don't give up Hattie. I think someday we will get the respect we deserve; I know you do good work. Even Emory said you were very good. Of course," she laughed, "he added, 'for a girl!'"

The attitude everyone seemed to have was the same one Mabel expressed: "You'll get married one day, Hattie, and you can get your husband to OK your work. I know it's not fair, but remember, we don't even get to vote!" She hit my sore spot! "Yet," she added. Not seeing the look on my face, she had no idea how very desperate I felt, as I had made up my mind never to get married.

I had decided this on a very first date with a very nice young man from the Methodist Church. His name was Clarence and he had asked me to go to the picnic, which was the big affair in Salem. I said I would go, mainly because Mama said that I should, because people talked about how strange it was that a pretty girl like me never had a social life. "It seems that people are asking me all the time why all you do is work. Doesn't Hattie want to be included in the young people's circle of friends?" I didn't want Mama to have to apologize for me anymore.

So to stop Mama from worrying about me, I said to Clarence, "Yes, I'll go and will bring 'our' picnic lunch." I knew that was the women's role, to pack a nice lunch for two. I was not very good with fixing food so I went over to Nellie's and asked her to help me with the lunch. She said "Of course I'll help you. I'm so glad you are going to the picnic with a young man. Do I know him?" When I told her it was Clarence Dyer, she said "Oh, I know his sister, Anna. They are a very nice family,

HATTIE

and they come to our church every Sunday." So as I helped Nellie with her young sons (she had two now, and was expecting a third child), she began fixing the lunch for me.

I felt that feeling of awkwardness, which came over me from time to time, in Nellie's kitchen that day. How different I was from Nellie, I thought. She seemed to know where everything was in this, her domain. Opening a large bread box, she took out a loaf of bread, and cut four slices. Then, she took some salami hanging from the ceiling and began slicing very thin pieces. I could smell the spicy sausage; it made me hungry. Nellie paused and looked at me smiling. She handed me the last slice of the meat she had just cut. I took it gratefully and said "Thanks, Nellie, this sure smells good." She took another for herself. We stood there enjoying the meat and the moment together.

. Nellie said "Hattie, I don't want to meddle in your business, but I have been waiting to talk with you. It seems all you care about is the newspaper. I know many young men who have shown an interest in you, and you know, my dear, we don't stay young and pretty forever. You can have the pick of men now. Find a good man, Hattie, as I have. Dan has property and a good job, no bad habits and he loves children, as I do. So, my dear, think seriously about Clarence Dyer. I'm so glad you are going with a young man."

I felt like unburdening myself, so I said, "Nellie I know you are right, I don't know what's wrong with me. Even though I know Clarence is a nice chap, I dread going on this picnic with him. He looks at me as if I'm an idol. He seems to just adore me, wants to hold my hand and sit as close to me as possible, and Nellie, I just hate his worshiping of me. I feel like running when he touches my hand." There, I told her my feelings, I had been looking down at the well-scrubbed floor in her kitchen as I was talking. Now I kind of quickly glanced at her face and saw she was looking at me with a troubled expression.

I knew I had made her uncomfortable with my description of my feelings, so I said making myself smile, "But I will try not to act foolish or insensitive." And I laughed. Nellie sighed, putting her arm around me. She said, "Hattie, don't be afraid of the physical part of

SHE WAS WIRED DIFFERENTLY

the marriage; it is really not bad." I almost laughed, remembering my experience with smoking and Half Moon, which was pleasant. For just a moment I thought perhaps I should tell Nellie. I wondered if this had something to do with my feelings about intimacy with men, but decided to hold my tongue.

I dressed myself in a soft summer dress and wore only sandals, because it was hot. "Papa said you could fry an egg on a rock outside in the sunshine," I said to Clarence to make conversation. I had brought the basket lunch that Nellie had fixed. Clarence took it from my arm as we walked to the church. He was tall and blonde and very handsome.

He had tried to hold my hand several times while we were walking but I really didn't want him to touch me, so I kept a distance between us as we walked. I asked myself, "What is the matter with me? What am I afraid of?" His hand was not clammy or wet like so many. No, I thought, it is something else, but I could not put my finger on that something. He was a quiet young man and had a good job working for the Great Northwestern Railroad, the same railroad that Dan, Nellie's husband, worked for. His work was hard, but he liked it, he said.

"I heard your father sold the newspaper recently, Hattie. What are you going to do? Seems that is where you have been spending most of your time." We were silently walking down the small dirt road toward the tree-shaded park area where the picnic was held.

"Yes, Papa did sell," I replied, and then feeling bad that I gave him such a short answer, felt I should tell him more of my plans.

"I want to travel. I want to go to Chicago first, and hope to find a job on a newspaper."

I asked just to be polite, not because I really cared, "What are your plans for the future Clarence?

He said, "Oh, I guess I will settling down right here, have a family, much like my folks, and their folks have done. Salem is a nice little town to do that in, don't you think?" I said, "If that's what you want, Clarence, I agree this is a nice town to raise a family in, just like my sister Nellie is doing, but I know that this kind of a life is not for me!"

I knew I was saying this with such strong feelings, almost scaring

HATTIE

myself with my own strong convictions. I felt surprised at myself for expressing this definite decision for my lifetime goals to him so forcefully, but I knew that I was letting him know that he would have no place in my life, now or later.

He gave up on me after that talk, I could tell. I can't remember the rest of the day but I know I was glad it was over. This is what I wrote in my diary: "I am glad this day is over, and he didn't ask me for another date, thank goodness. I hope Mama won't be too disappointed. She is still stuck with me!"

We stayed in Salem until 1898. Papa and Mama were forty-seven years old and I had turned twenty-two; Fred was eighteen; Lizzie seventeen and Ethel was seven. Mama had wanted Papa to build us a house in Salem but he needed to take up carpentry again.

"The newspaper business could not pay the bills," I heard Papa tell Mama. "I need to do carpenter work, Liz. That means bidding contracts as you know, which means we will have to move where the work is."

I was beginning to understand why we had always needed to move about so much. Poor Papa was just trying to keep a roof over our heads and food on the table. He bid on a schoolhouse in Mitchel, South Dakota. His bid was accepted and so we needed to move this time to Mitchel.

The Spanish American War had begun. Fred was eighteen and he decided to go into the service, and into this war, rather than move with us to Mitchel. I did not think this was a good idea, but he wanted the adventure, and I understood that.

Papa, Mama, the girls, and I said "goodbye" at the Salem train station as Fred left South Dakota heading for Colorado. Mama gave him the basket of food that she had prepared so lovingly. She cried over the fried chicken as she was cooking it, saying,

"This is my dear boy's favorite food." Then she put in several pieces of cake wrapped carefully in napkins. The tears dripped down onto the napkins, dotting them with little puddles of water.

She sobbed, "My dear boy, going off to war, will I see him again?"

SHE WAS WIRED DIFFERENTLY

She was asking herself and looking up to heaven for an answer.

I said, "Mama, he is eighteen now, not a baby, and he is big and strong; he can take care of himself. "

But on that station platform watching him board the train he looked so young. My heart was in my throat, and I yelled at him as his head popped out from one of the train's windows, "Promise to write to Mama and Me!" He yelled back, "I will!" We had hugged each other over and over, and I felt myself wanting to cry as his train pulled out. When would we see him again, I wondered.

CHAPTER SEVEN

Dakota weather is stifling hot in summer and bitter cold in winter. People adapt in the summer by getting most of their work done in the early morning. In the winter, they shovel snow all morning cutting firewood, keeping a stack near the fireplace for the cold nights.

Mama had found a job cooking for a small café' in town. She walked to the café before sunup to make the coffee and get things ready for the early-morning customers. Papa worked hard that first year in Mitchel building the schoolhouse he contracted for.

Soon after he finished, he left to prove up on a claim for some property up north. He had his eye on a property that he thought would bring a "nice windfall," he said. He had his name on several government land patents in countries all over South and North Dakota.

I got a job in a printing office in Mitchel making three dollars a week working nine to ten hours a day, six days a week. I was doing all the typesetting that the owner's wife had been doing. She was expecting a baby and was sickly, so I was to do her work until she was able to come back.

I gave Mama two dollars a week and saved fifty cents for my dream trip to Chicago and used fifty cents for my expenses. We had found a small apartment over a bakery. It was a bit crowded. Mama and Papa had the front bedroom with a large bay window overlooking the street.

HATTIE

Mama had placed her old rocking chair in front of the window where I was allowed to sit; I pulled back the old heavy lace curtains while I read or wrote in my diary. I loved to sit and rock looking at the street below, thinking, dreaming, and planning my move to Chicago.

The back bedroom was small. There was just enough room for Elizabeth and Ethel in a full bed and for my cot pushed against the other wall, leaving a narrow walkway between the beds. A small table stood under the window overlooking an alley running behind the main street We girls heard the banging of the downstairs bakery door every morning as the milk was delivered. We knew exactly when the baker, his wife, and children all came to work each day. They were very noisy, but I loved the smell of the freshly made bread floating up from below soon after their arrival. This made me ravenous and I jumped out of bed to fix breakfast for the girls and me.

Our large dinning room had a big, round, heavy table in front of a fireplace that kept the whole place nicely warm. We spent most of our time sitting here eating, talking, sewing, and playing games.

Lizzie was eighteen and seemed very dependent on Mama. She graduated from high school and wanted to go to college. She needed to find a college, Mama kept telling Papa, where she could work for board and room, but this was hard to find. He thought she should be getting a job to pay room and board for herself. Mama, always a little soft about any of us children, allowed Lizzie to follow her interests and encouraged her to find a college.

I heard Papa say, "Elizabeth, you are allowing her to cling too close to you. She needs to get out into the world and make a living for herself. She is scared of her own shadow." Mama argued, "She is weak when it comes to meeting people. She has always been that way, ever since we lived among the Indians. She became frightened back then; Melroy and I want to help her to get over it. Maybe her wanting to go to college is her way of getting out in the world."

Papa threw up his hands and said, "Have it your way, Liz, but I think you are babying her too much. We need the money she could be making. I think she should see this herself. Here you are, working for three dollars and seventy-five cents a week, Hattie working for three dollars a week, and me maybe making ten dollars a week and she is making nothing at eighteen. That's just not right."

I agreed with Papa, but I didn't say this to Mama because she was so worried about Lizzie. Also, Ethel was very attached to Lizzie and Mama felt if Lizzie would go away somewhere to work that would leave Ethel alone. Ethel was only thirteen and with all of us working it seemed like a good thing for Lizzie to be home with her. I knew Mama was hoping Liz could find a college close by. Poor Mama, I thought once again, she has had such a hard life and has had to fight for everything. I determined, there and then, that I would continue to help her with money so she wouldn't have to worry so much.

I spent many hours thinking and studying about how different men and women's lives were. I continued to see women's role and men's role in life more clearly. At first it puzzled me and then began to make me angry. I found myself turning against the church and the Bible because I had come to the conclusion that it was God's fault. He had let women down.

I watched as my two older sisters accepted their place of being a passive partner. They seemed eager to want children and the home life that I found discouraging and unfulfilling. I kept asking myself why they accepted this uninteresting role in life. Why, women couldn't even vote! It was not fair, I said over and over again. Why am I so different than other women, I wondered.

I was beginning to give serious thought about when I could go to Chicago. How to accomplish that was always on my mind. I collected every bit of information I could find about the city and the Chicago newspapers. I was interested mainly in the Chicago Tribune, where I wanted to work. I stored every scrap of information about that newspaper and kept a file that got bigger and bigger.

Papa spent more and more time away from us. He was drinking pretty heavily again. Many nights he did not come home. He was building stores and homes, even a school during our first few years in Mitchel. He was moving from town to town and from South to North

HATTIE

Dakota. I figured he just went to the building site, drank, and slept off the drink. I remember him saying, "Well girls, I'm going out to prove up on another claim." And off he would go and we wouldn't know when we would see him again. Mama would get a few dollars in the mail every week or so, and I could tell by her sad eyes, when she opened the envelopes and pulled out the money (which never seemed to be enough to cover our expenses), that she wished Papa were here with us.

When Papa began drinking and staying away, Mama began confiding in me as she had done with Nellie. She said little things like, "I don't understand your Papa falling in with that bunch who hangs out at the saloons. I wish he would come to church with us." Mama sighed. His absence made it hard on her; she never knew when he was coming home or how much money he would bring when he did.

I didn't want to talk about Papa to Mama, because he had such a special place in my heart, but the older I got the more I could understand Mama's feelings. Again, I blamed God. He allowed Papa and the other men to forget their worries and troubles through whiskey and gaming and with other women. I had heard about it. I couldn't believe that "our" Papa was doing "those" things, but many men were, and I considered it shocking and unforgivable.

I went to church with Mama and the girls every Sunday for Mama's sake. I would listen to the sermon. I liked the ones best that were about being a good person and how to love. But when the preacher would preach his interpretation of Jesus' words, I didn't agree always. I would sometimes talk to Mama about my feelings, but most of the time, she did not understand my arguments and would seem to get upset by my questioning the preacher, so I began to just put my thoughts into my diary and not bother her with them anymore.

Mama had only the church to help her; it seemed that all the women that attended church had problems. I wrote in my diaries and I had stacks of them now. They filled the bottom of my cedar chest.

Every young girl had a cedar chest. Most filled theirs with pretty embroidered pillow cases and hand towels. Papa had made Nellie, Mabel, and me a gift of a beautiful "hope chest," as they were called,

SHE WAS WIRED DIFFERENTLY

when we each turned sixteen. Nellie and Mabel had spent many hours going over all the items they had carefully placed in their chests. And as I never planned on getting married, I was filling mine with my diaries and my beloved books and a few Indian articles I had saved. I talked to my diaries like a best friend. Sometimes it was like praying. I so wanted some way of figuring out life.

I also began writing essays. They were my way of expressing my feelings. I tried poems. I wrote one to Mama for her birthday in 1897, just before we moved to Mitchel, when she was forty-six years old.

> HATTIE COMPOSED THIS HERSELF (Mama wrote)
> My Birthday Offering To You
> By Hattie Fuller Age 21
>
> Today the day of your birth
> And life is full of love and mirth
> Forty-six years have over you fled
> Going swiftly with noiseless tread.
> Forty-six years in this world of care
> And grief's--you've had your share
> We hope as many more may flee
> And you will ever happy be.
> The joys of life are coming yet
> And you will have them, don't fret
> You have some girls as big as you
> And grandchildren, you have too
> And when on earth your life is done
> And a place in heaven at last earned
> And then repeat. I lay me down
> At last you've won a priceless crown.
> From your loving daughter

Lizzie found a college that she could afford. It was in Owatonna, Minnesota. She was to wait on tables and clean classrooms to earn her

HATTIE

tuition, room and board.

After Lizzie left for college, Mama and Ethel moved to Spring Valley, Minnesota, into Mama's old home where her mother was living in her big two-story house in the country. Some of Mama's sisters and brothers still lived there with her, after their father had passed. Her children kept house for her and were taking care of the land, farming it like their father had done.

After Mama had gone back to Spring Valley, I found a room with a family in Mitchel. I paid them a dollar a week for room and board, saving a dollar for myself. I sent Mama fifty cents a week. She didn't need as much now that she had no rent to pay; that left me with fifty cents to spend on myself, and I managed very well.

I had made much money for the print shop in Mitchel. And when Mr. Dennison's wife came back to work they decided to keep me on. I had taken a couple of Papa's tips on how to improve circulation and was able to build up the readership a little. The first thing I did was visit the principal of the school and ask him if I could interview some senior students about what they wanted to do with their lives after they graduated, and I would write an article about them.

The principal thought the idea was a good one. It created a great interest among the parents and the families of the students. Papa used to say how much people loved to see their names in print, and then he would laugh and say, "Anyone whose name you put in the paper will have to buy one or two extra copies to send to family and friends." He would laugh and do his little shuffle dance.

So when I did the interviews with the students, sure enough our paper sales went up. Just as Papa said, the students' families would buy a couple of extra copies.

Another of Papa's suggestions was to circulate a flyer advertising a handbill for businesses in town, describing what the businesses had to offer. The flyers would be delivered with the newspaper. The businesses always bought a few extra copies as this stimulated more sales.

I worked for Mr. And Mrs. Dennison for about four years and had saved over $150. This is the amount of money I felt I needed to move

SHE WAS WIRED DIFFERENTLY

to Chicago. I put this money in the bank.

I saved enough money by 1903 to pay for train tickets for Mama and Ethel to make the trip with me. "Mama," I kept pleading with her. "Please come and have a vacation, with me. See the city and maybe a play." I added, "It will be good for Ethel to see another part of the world. Then if I do find work, you and Ethel can travel back together." This was my dream and of course Mama knew it.

Finally she agreed to come with me. I think she did it for Ethel as much as for herself. Ethel had been lost without Elizabeth and seemed excited about the idea of a trip to Chicago. So we began planning and packing the clothes we were going to take running around Grandma Taylor's house. Grandma got excited for us. "Oh, my goodness, you three are acting like children. Do slow down or you are going to fall down the stairs," she kept saying with her very proper English accent.

I tried on all my new clothes for the tenth time. I had been buying my wardrobe piece by piece from the Sears & Roebuck Catalog. I had already packed a big trunk and my hope chest, giving instructions to Mama about when to send them to me.

The trunk and chest were precious to me; they contained my collection of diaries, my books that I just could not part with, and all my clothes, bedding—everything I owned in the world. Mama said, "Hattie, please darling, don't worry, I will see to it that it will be sent to you. Do calm down; you are running around like a chicken with its head cut off," she laughed. I was glad to hear her laugh; I hadn't heard her laughter in such a long time. I knew she missed Papa, but I knew also that the two of them were not very happy when they were together anymore. This made me sad, but I didn't know how to fix that.

I looked at myself in a long mirror in Grandma's room. The woman who looked back was a stranger. I had only seen myself in a little mirror over my sink in my room in Mitchel. I thought I looked good except for my serious expression. Nellie and Mabel had called me "Grandie" at times; now I knew why. I had a solemn, old-lady look. A very serious, stern-looking young lady was looking back at me. I turned around, and peeking over my shoulder I could see that I was not too tall—about five feet six inches—and I was slim, with a small waist. I always wished I was taller and stronger looking, but it wasn't to be.

I picked up a pair of eyeglasses that I had bought, thinking they would make me look older and more sophisticated. I put them on and looked again. Yes, I told myself, I was right: this was the perfect touch. The glasses made me look the studious type; it wouldn't pay to look too young or too pretty for, the job I was hoping to get.

Mama came into the room and said, "Oh, Hattie, you look perfectly elegant. What a lovely suit." I turned to look at this dear mother of mine, hearing in her voice the wish that she too could look elegant again. "Mama" I said to her, "let's look in the Sears Catalog and find you and Ethel something elegant to wear too." An excitement and little girl look came over her face. So off we went to find the catalog.

We had so much fun looking at all the clothes. I thought Ethel was going to cry when I told her she could pick whatever she wanted. Mama warned, "But it has to be sensible, Ethel!" They both chose two-piece suits, with white lace blouses, that I knew would make them look just right for a vacation to the big city.

I then walked to the photographers and had my picture taken in the big hat and glasses, as I had promised Mama.

The day finally arrived for our departure. As we got on the train I hardly looked back. I was not sorry I was leaving. Papa had finally come to see us off and gave Mama some dearly needed money. He said he was building a newspaper office in North Dakota and was still taking jobs for carpenter work. He gave Mama a big hug and told her that soon as he finished this job he would send her more money. He was looking kind of tacky. Mama looked at him with sad eyes. He hugged Ethel. She was almost as big as I was now, and she looked a lot like Mama, I thought.

Then Papa turned to me, and reaching for his inside breast pocket, he pulled out two envelopes and handed them to me. I took them and looked at his handwriting on the first one, which read, "To Whom It Might Concern." I looked at the second and across the envelope was the name Thomas A. Lewis, Editor. I looked at Papa for an explanation. He smiled at me kindly, and said, "This letter is to an old friend in the newspaper business in Chicago. Try to look him up. The other is an introduction for my daughter and a fine newspaper person who I have taught personally her whole life. It tells of my newspaper experience and the papers I have owned all over the Midwest. I made a list of the papers and their locations. They make a dandy list," he said with a chuckle. "I told them that if given a chance, you would do a very good job for them as you had been taught very well by me."

I was so touched by his care and love for me. I knew I would treasure this moment and these letters for the rest of my life. I gave him a big hug and kiss, which was not my usual way. I gave him one of the pictures of me in my new suit with the big hat and the new glasses. He looked at it and held it to his heart with tears in his blue eyes. He smiled that sweet sweet smile of his. I said, "I'll love you forever Papa." And he replied, "Me too my little Chickadee. Take care of your Mama and Ethel and have a good time."

"Goodbye Papa, goodbye," I whispered to him as I hugged him again. I wondered if I would see him again. I knew I would miss him; I already did.

Once on the train, I heard that long, soulful whistle and I felt us moving. I looked at Mama, and she looked pleased. I think for the first time in many years she felt the worries of the world fall from her shoulders. I was so glad I had coaxed her into coming and I was determined that she enjoy herself. I had everything planned. I would see to it that Mama and Ethel would have a good time, and then I would start looking for work.

Hattie Age 26

The Wedding
Inez and "Cappie" (Hattie Fuller)

CHAPTER EIGHT

I had read of faraway places in books, and heard Papa and Mama speak of their travels. But until we were actually on the train I found that I had little concept of great distances. The beautiful, spring landscape rolled by as the train made its way through South Dakota into Illinois and finally arrived in Chicago, the biggest city in America outside of New York. I was spellbound. I couldn't believe my eyes; there was so much to see.

We were expecting a big city, but we did not know what a big city looked like in 1903. Astounded, we watched out the window as our train approached Chicago. First we passed small, scatted farms. They became closer and closer together as we neared the city. The farms gave way to factories belching plumes of dark smoke. We had our noses glued to the train window. The landscape changed first to big red brick buildings, then to taller and taller buildings. Soon they towered over us on both sides of the train. Buildings so tall we could not see the tops of them. It was breathtaking.

Finally the train pulled into the station with its whistle blowing as it began slowing down. We arrived at a long platform. The train settled down and released a gasp of steam, the signal that we had reached our destination. We clutched our bags and began hanging on to each other as we made our way slowly down the narrow aisle, terrified of losing each other. Finally we stepped off the train onto the long platform.

It seemed like hundreds of people were rushing toward the train and at the same time, people were pushing us from behind to get off. The hustle and the noise surprised me. I felt my heart pounding as I was being pushed this way and that. I worried about Mama and Ethel, but saw Mama had a strong grip on Ethel as she clutched my arm.

I felt immediately in charge. I didn't want Mama to be frightened, so I plucked up my courage and yelled, "Come Mama, this way, we will find a porter! Hold on Ethel, tightly!" And off I went with this new attitude of someone who knew exactly what she was doing. For a few moments I felt like an actor playing a part. As I moved through the crowd, I became more and more confident. I felt an air of boldness and realized I was really not a bit afraid.

The chaos was exciting and I smiled to myself. I held the door to the train station open for Mama when a gentleman reached over my head. He grabbed the door, holding it open for Mama, Ethel, and me. I looked at his smiling face and mumbled a low "Thank you, sir." He tipped his hat and said, "A pleasure, miss." I felt myself go back into the shell. "What was this," I thought. "What is this thing inside of me that makes me shut down when a man shows any kind of interest in me?" I must think about this.

We were standing in a beautiful train station with our mouths open looking at the tallest room we had ever been in. I finally stopped gawking and remembered my plan.

I found a porter who gathered our bags up and he found a carriage to take us to the hotel, where I had booked our rooms. Mama seemed to walk a little straighter and with more dignity than I remembered and even Ethel seemed more grown up. I was proud of us.

Soon we arrived at the "sweet little hotel" that Mrs. Dennison had described to me when she heard I was going to Chicago, her hometown. She had said, "It's not the Waldorf Astoria but it's very nice and clean. They will feed you breakfast, but best of all, you can walk to almost all the stage plays, and window-shop all day in grand shops on the Loop. We stopped in front of a nice-looking, stone-front, three-story building. I paid the driver and noticed the area seemed very proper and the hotel appeared to be clean. I will write and thank Mrs. Dennison, I thought.

The small lobby was clean. The only other hotel we had stayed in was in Salem, and that was more like a boarding house. This looked pretty fancy. A young man was standing behind the counter looking at us with a broad smile on his face. I marched up to him and announced in a rather loud voice, "You have a reservation for the Fuller Family." I startled myself, so I lowered my voice and said, "We are the Fullers."

He looked us over and opening a large book in front of him, said, "Yes, we have you in a front room." He tapped a bell on the counter. "Sam will take you up in the elevator."

We had never been in an elevator before. Sam seemed to know this and that made me feel like a hick all of a sudden. He seemed to be looking at us with a superior tip to his head making me feel ignorant, because he could tell.

He looked like the young clerk's brother; we found out later he was. His family ran this small hotel. His grandparents were the original owners. We all became friends during our stay, but at this moment I was trying to keep this kid from getting under my skin.

Mama saw my displeasure and as always smoothed the way. "You are a very smart young man to know how to run this elevator" she said, in her most modest voice. He looked at her proudly and replied, "I've been doing this for a whole year now." She said, "You are a very smart young man."

This simple discussion broke the ice. He had been eyeing pretty Ethel from the moment he set eyes on her and surely saw Mama's friendliness as a way to get to know her. So, taking advantage of the situation, I asked as we stopped on the third floor, "Do you know where *The Rose* is playing?"

"Sure I do. It's only a few blocks from here. If you want, I can show you. I get off at four o'clock."

"That would be wonderful," I replied sweetly, as he opened the door to a nice large room overlooking the street below.

HATTIE

Mama said, "That would be very nice of you. Your name is Sam isn't it?" He looked bashful all of a sudden, and very young.

"Yeah," he answered. Mama then spoke as if she were at a church gathering. "Well Sam, this is my daughter Ethel and my older daughter Hattie." He looked kind of embarrassed as he shook our hands.

"What a nice room," Mama said. "It is one of our best," he replied with pride. After he left, I felt that we had, indeed, met a friend, thanks to Mama.

We saw several plays and took in most of the sights around Chicago. Lake Michigan was the biggest body of water that Ethel and I had ever seen. Of course, Mama remembered when she was seven years old and had sailed on the ship from England. She told us the story once again as we sat on a bench bundled up in our warmest coats and scarves, overlooking the lake. It was a nice day but the wind was cold blowing off that big, grey body of water. We watched the ships moving slowly up and down the lake, but it was the people we enjoyed watching the most. They were walking, feeding the birds, and stopping to gaze at the ships and small boats along the footpath.

We went to a large, world-famous museum and rode all over town in horse-drawn carriages for very little money. There were so many tall, beautiful buildings; we got stiff necks trying to see the tops of them. Almost all the carriage drivers reported the ones that were world famous, pointing them out to us proudly. I had never seen Mama so happy and so rested. Ethel too was enjoying herself; she didn't even talk about missing Liz except once when she said she wished Liz could see the sights too.

We heard Enrico Caruso's voice on a new gramophone recording while in one of the lovely department stores on the Loop. This area in Chicago had many stores, theaters, and restaurants. We spent lots of our time gazing in the windows at all the beautiful things for sale. We had heard many stories from travelers coming through South Dakota, but it was different to really see and experience the city itself.

The last day before they had to go, Mama and I talked about all she had seen in her lifetime. She said these last two weeks were two of the most enjoyable she could remember. I was so pleased. We had eaten and were spending our last evening in the quiet of our rooms; all the packing was done. "Hattie," she said looking up at me as I stood over her chair, "I know Chicago is an exciting place for a young woman, but it is so busy. Everyone seems to be in such a rush. What if you get sick? You know no one. Who will care for you if you get sick? Do come home with us. We will miss you."

I looked down into her concerned eyes, knowing she loved me dearly. "Mama, I will be fine." She could not understand me. "When you were young, you had too much willpower. I always knew if you wanted something, Hattie, you were like a dog with a bone; you just never gave up. I hope you will find what you want in this big city. You know no one, and no one knows you, but I will keep you in my prayers, my dear." She sighed patting my hand, feeling this was all she could do for this headstrong daughter of hers.

I put my arm around her slightly plump shoulders, and said, "I know you are right, I do have lots of determination. I think I get it from Papa." I laughed. "But I'm glad I have it. I don't see anything wrong with this attitude. If I think something should be a certain way, then why shouldn't I try to live my life accordingly? I am proud that I have developed certain principles to live by.

"Please don't worry Mama; I'm going to be just fine. I will write you, and tell you what I am doing. I've given this a lot of thought. Just keep praying for me that I will find a newspaper job right away. I will send you some money, as soon as I can. I love you, Mama. You have worked so hard your whole life. I hope you can take it easy from now on. I'm looking forward to my life here in this big city. I am very excited."

The Benson family, which owned the hotel, had taken a liking to me. After Mama and Ethyl left for home, Carol, as I began to call Mrs. Benson, offered me a job setting up the tables and serving breakfast. They served breakfast to people who stayed in the hotel, but also had opened a small breakfast room off the lobby for the general public. It was beginning to get busy. So when Carol said she was looking for

HATTIE

someone to help serve breakfast I asked for the job. After hearing of my plans to stay in Chicago, she hired me. I was delighted. I felt this job would give me time to find the job of my dreams. I had saved quite a bit of money for this move, but it was going fast, so the job was just perfect.

After serving breakfast each morning, I ran the two blocks to my room in a very old-fashioned home that belonged to an old stage actress and her actor husband. Carol again had come to my rescue. These French people were friends of her family. They had known one another in France. Carol knew they wanted to rent a room in their large home, so she had recommended me to them. They had a daughter, Inez. She was about my age. They all spoke with a French accent that I found very enchanting. They were fascinating people. The more I got to know them, the better I liked them. I had never met people who were so open-minded about everything in life and laughed so much. It was captivating to me; their lifestyle seemed so open and warm.

Inez and I became strongly attractive to each other. She had such charm. I was transfixed by her and could not get enough of her company. We began spending every extra hour with one another. She seemed to like me as much as I liked her. We began taking long walks together, sometimes along the Great Lakes shores, sometimes looking in shop windows. We talked and talked and the days flew by. She knew so much about plays, actors, and books that I had never heard of—mostly fiction. I was interested in politics and American history and in women's causes. She became interested in these causes too, maybe because this is about all I talked about.

I had been following Susan B. Anthony. She had published *The Revolution,* a magazine that was centered on the Women's Suffrage Association. Their motto was: "The true republic-men, their rights and nothing more; women their rights and nothing less." Susan B. Anthony had become the acknowledged spokeswoman for the campaign. She was a spinster devoting her life to the women's movement.

I told Inez all about this hero of mine. At 82, Anthony had been born in 1820, in Massachusetts, but she was still going strong. It was

SHE WAS WIRED DIFFERENTLY

said that Anthony, even as a young girl, had an independent spirit. I told Inez one of the stories I had read about her as we were sitting on a bench one beautiful chilly Chicago day.

We both laughed as I related the story. Anthony had asked her male schoolteacher why he only taught long division to boys. Anthony was not satisfied with his answer, which was: "A girl needs to know how to read her Bible and count her egg money, nothing more." So, Susan reorganized herself around the teacher and sat behind him in order to learn long division. Susan demonstrated her strong strength and will, even as a young school girl. She learned long division!

Inez was enthralled with that story, so I related the story about Susan B. registering and voting in the election of 1871, and getting fined.

I was so glad to find someone to talk to about my inner thoughts, which up to now I had only written about in my diary. Inez told me about the plays in which she had bit parts. She had just landed a stand-in role for the part of the lead actress and was excited. She laughingly said, "I pray every night, Hattie, that she really does break her leg. Isn't that sinful of me?" We both laughed. I said, looking at her pretty face, "I just know you will get your break. I feel I will find a job doing what I want to do." I reached over and squeezed her hand and felt a tingle go through me.

We became inseparable. She liked to be with me and loved it when I came to the theater to watch her in a play, even though I had to sit in the balcony to save money. We would go out for coffee while she still had her stage make-up on. Everyone in the small café that we frequented knew she was an actress and showed her respect and a bit of awe.

I finally landed a job with a printer. He was an old man by the name of Mr. Hatch. He said while interviewing me, "I'm an old man and pretty tired of this business, but I don't want to sell it yet, so I think I'll hire you to do all the stuff I don't want to do. All the dirty work," he laughed. "You look like a smart girl. So I won't have to tell you a dozen times the same thing, now will I?" "No, sir!" I answered. He reminded me of Papa and that made me feel very comfortable. I was excited. The

HATTIE

pay was $5 a week, but he said, "If you show yourself to be worth more, I'll pay you more." And that was how I started.

I took Inez out for dinner that evening. We both got dressed up to the hilt. Even though it was raining a little coming home, we laughed and I held her hand as we ran through the rain. I had been in Chicago almost six months and thanked God that I had the waitress job, as it was paying my expenses and a little extra. Also, I was able to keep it and get to my new job on time.

We got back to the house and in the darkened hallway she turned to go into her room. I started to turn to go into mine when she threw her arms around me and whispered, "I had the best time of my life with you tonight, Cappie." That was her name for me now. I looked at her upturned face. Her soft pink lips were so inviting so I leaned down and kissed her on that pretty mouth. It seemed the most natural thing to do at that moment, and she responded to the kiss by giving me a little flick of her tongue on my lips as we parted.

She did not seem surprised. In fact she seemed pleased and gave me a long look of complete open attraction, like a plea for another, as she leaned against me. She was a bit smaller than me. I put my arms around her again and drew her back into my body; I felt at that moment that all in my world was finally right. I finally knew that this was the feeling people felt when they were in love. I never wanted her to leave my arms.

We stood like that pressed together in that dimly lit hallway for it seemed like forever; her head was on my shoulder. I could feel her breath on my neck. She smelled of gardenias. I knew I would never forget this moment of enchantment. She stirred and began to move away. I let her move but only a little, just enough to look down into her eyes again, to see if I could see that same look of attraction that I had seen a moment before. Was it still there? Yes, it was in her eyes, and on her soft lips. She had a half smile, like a cat that had caught the mouse. Did she know about these feelings all along, I wondered, as I leaned down and kissed her again and then again. Until she pulled back and said in a whisper, "Enough, Cappie, enough for now, dear heart."

SHE WAS WIRED DIFFERENTLY

I could not believe what had just happened to me out there in the hall as I lay in bed that night, but I knew it was just the beginning of something wonderful for me. Something I had waited a long time to happen to me. I had seen it happening to others. It was like an awakening. I was so happy, I couldn't believe my good luck in finding Inez. I couldn't wait for tomorrow. I couldn't wait to hold her and kiss her again.

CHAPTER NINE

In the first year I worked for Mr. Hatch, I found he was a wonderful man, much like Papa. He treated me with kindness and dignity; he taught me about printing books, which was his main business. He printed the most beautiful books. The covers were of fine leather, the paper was made from light linen, and the typefaces were unique. I learned about different kinds of paper and how to put pictures in books with a cover of tissue paper to protect the picture. He taught me about bindings and glues, about how to arrange the sections, about indexes and headers and footers, so many things I had never known. He sat on a high stool as he overlooked everything I did. I was happy to learn, and he loved to teach, saying at the end of each day, "You are a smart one, Hattie." "You learn, fast! I like smart people."

I began to wear men's shirts and a tie to work. Many women were dressing this way, and one day I cut my hair into a bob, this also was the new style. Mr. Hatch didn't seem to care or notice, but he soon began calling me "Boy."

One payday he raised my salary by a dollar a week without telling me. He just added the dollar in my pay envelope. I thought he had made a mistake and tried to give it back to him. "Mr. Hatch, you have overpaid me a dollar," I said.

He just looked at me with a sly grin and said, "And, you're honest

HATTIE

besides being smart. I was really in luck when I hired you! No, boy, it's a raise in your pay that I promised if you proved out to be worth more money." I immediately gave up my morning waitress job and sent Mama a little extra.

I got letters from Papa, keeping me up to date on the family doings. I wrote telling him I was learning to bind and print books. I also told Papa about Inez, and how much we liked each other. I didn't tell him about the kissing and fondling that was going on between the two of us, of course, but I told him I had found the love of my life and left it at that.

I also wrote to Fred and explained my feelings for Inez. I felt he knew me better than anyone else in the family. I reminded him of how I used to act, wishing I were a boy. I told him that I had begun making drastic changes in my life in Chicago. I started wearing men's clothes and acting like a man.

I needed to tell Fred what I was doing and explain my future plans to him. I was going to use Papa's middle name, Alfred, I told him. I planned on using Fred's birth date and birthplace unless I heard from him that he definitely did not want me to. I felt confident that if I tried for a job on a paper dressed as a man, I could get hired. I knew as a woman I did not have a chance. "Please help me, little brother," I pleaded, "Please tell the rest of the family of my plans."

In my heart I was sure about what I was doing, but I believed the family might take it better if the news came from Fred. He could remind them of how I was when we were young, and how believable I was when dressed as a boy. I just didn't have the courage to come out with it myself to my sisters and Mama and Papa. I thought if they found out this way, through Fred, it would not be such a blow. That was, if they ever wanted to see me again, which sometimes I doubted they would. I knew this letter was burning my bridges behind me, but I was committed.

Sometime later I read Mama's letter to Lizzie in Owatonna. It said: "I got a letter from Nellie; she copied Fred's letter to her, and sent it to me. So I copied it and am sending it to you to read, but be sure and

SHE WAS WIRED DIFFERENTLY

return it as it will save me copying it again, and I want to send it to the others to read. It just made me sick to read it, but be sure and not tell anyone about what Hattie is planning to do so it will not get in the newspapers. It could make trouble for Fred and for all of us. Please just wait till Fred gets home and then we can all talk. I will let Ethel tell you all the other news. Goodbye and God bless and keep you. We both send love from your loving mamma and sister."

The next thing I heard sent me into a deep shock. Mama died on November 2, 1903; she was only fifty-two years old. She died in Grandmother Taylor's home in Spring Valley, Minnesota, after a very brief illness. This was all Elizabeth said in a letter to me. Papa didn't seem to know what had happened and said he would try to find out more and would write.

Grief and guilt overtook me. I couldn't work or sleep. I worried if my news to Fred, which he had written to Nellie who sent it on to Mama, had been so frightful it had caused her to have a heart attack. What had I done? I asked this question over and over again. Had Mama not been able to understand this need of mine? Was she so afraid of me being found out that it scared her to death? She knew I had always been different from my sisters, but she didn't know just how different, even I didn't know how different I was before I had met Inez.

Inez and I talked about our feelings. We both knew from a very young age that our feelings were not like those of our friends. Inez told me she had been first attracted to a woman, a friend of her mother's, when she was quite young. She had much more knowledge about our behavior than I had and went into detail about what she called the "him/her facto." We are not conventional people, she said, because somehow we are physically attracted to our own kind.

"Cappie, there are thousands of 'us' in the world. My Mama told me this when I fell in love with her friend." She was looking so young and pretty as she told me the story, while looking into my eyes. All of a sudden, I was terribly jealous of her long-ago love affair. A feeling of envy spread into every part of my body, washing over me like a bucket of water had been poured over my head. I wanted to grab her and hold

HATTIE

her to me forever! I became so afraid someone would take her from me. And all of a sudden I remembered Doe, the beautiful Indian girl who ran out of that bar on the reservation. Yes, this was the same feeling I had then. I wanted us, Inez and I, to be together forever!

I listened then to everything Inez told me about our difference. The explanation she had learned from her parents. Many stage people were like us, they told Inez. She met many while growing up. She saw no reason to feel ashamed. I wanted to believe her, but I knew that our relationship was not acceptable in most of America. I was confused and sad at times, and at other times I felt defiant and angry and always frightened at maybe being found out.

I was in love one moment and so depressed about Mama and our situation the next. Then I would see Inez. I would wait at our favorite meeting place, a bench in Central Park, and she would come running toward me, her small figure so perfect in its shape but bundled up in her winter coat—the collar up, a hat pulled down around her tiny ears. The closer she came to me the more I could see of her pretty face. My heart raced with joy at the very sight of her. How could this be wrong, I thought. I dreamed of her night and day.

We so enjoyed everything we did together, but most of all, we loved our time alone. When I was alone with her, her head on the pillow next to me, I would trace her pretty little face with my finger. It was like I was blind and wanted to remember its every intimate detail. I started at her eyebrows and very softly ran my finger over her brow and down the right side of her heart-shaped face to her chin, then outlining up the other cheek to her brow again. My finger would softly trace around her pink lips, slightly swollen from my kisses. All the while I would be looking into those deep blue eyes that sometimes seemed to turn green or gray, seemingly depending on the day or the weather. She laughed at me and said it was neither the day nor the weather. It was what she was wearing. I spent our time alone watching her, touching her, holding her, and telling her how much I loved her and adored her. She loved my attention and my admiration of her. How could our love be wrong, I asked myself day after day.

SHE WAS WIRED DIFFERENTLY

I was devastated about Mama. When I thought about her, I couldn't believe I would never see her again. Memories came crashing down on me. She had seemed healthy and happier than I had seen her for years; we had walked all over Chicago. And I relived that last night together when we had talked well into the night, like two friends rather than mother and daughter.

I was torn between depression and guilt. I shared my guilty feelings with Inez but it was hard for her to understand. Her parents had watched us fall in love. They did not seem to think it strange, because in the theater business they knew of many relationships like ours. They accepted us and were happy for Inez. "You are good for our daughter, Cappie," they said. "She never had a love before and we worried about her until you came into her life."

Inez, trying to be understanding, listened to me quietly and then said, "Cappie, you told me many times about what a hard life your mother had. She was probably just worn out, my dear heart; don't feel it is your fault. I'm sure she wanted nothing for you but your happiness." Her words helped me start to recover, but still, I was anxious to hear from Papa. I could not be completely at peace until I knew what caused her death and how Papa was taking the news.

I hadn't heard a word from Fred since my letter to him, nor had I heard from my sisters. Finally Papa wrote me that Fred was out of the service and was courting a lovely lady by the name of Nellie Rice. They were planning on getting married and had come to North Dakota looking around for some land near Papa's new newspaper business. "Won't it be funny that we will have another Nellie Fuller in our family, Cappie?" He addressed me by my nickname. This was a good sign I felt. "I still don't know what happened to your Mama," he wrote. "I have written to Elizabeth asking about her last days, but I still have not received an answer." Papa didn't say a word about my new personality. He had always just seemed to accept me. I was so grateful for his attitude.

Papa had bought a newspaper in North Dakota called *The Western Call*; he had been operating it for a little while. He wrote me shortly after Mama passed telling me that Ethel was traveling to live with Elizabeth. The letter had been a long, rambling one, first about Mama, then Fred, then Ethel, then back to Fred: "Fred had saved some money while in the service and wanted to buy a ranch to raise horses on. He is looking around my area in the Dakotas."

I knew that Fred had always dreamed of raising horses. I was happy for him. He had found a wife, and it looked as though he would be fulfilling his dream of a horse ranch. Should I write him, I wondered. No, I told myself; it was best to wait for him to contact me. I had not heard from any of my sisters, either. I knew that they were all busy with their families and probably didn't want to confront me about my decision. Papa was keeping me up to date on everyone and did not seem to care what anyone thought. He was the most nonjudgmental person I ever knew. How lucky I was to have my Papa, I thought.

I figured I just may never see the rest of my family again. It was hard to think of my sisters and their families, my nieces and nephews, and children I may never know. It was hard realizing that I may never see Fred or meet his wife. For the rest of my life! This was hard to think about. But I had made my decision to live my life honestly, and I hoped someday they would understand.

Inez and I walked everywhere that year. In winter we bundled up and walked through the parks and watched the ice skaters. One day Inez handed me a large package, "What's this?" I asked.

"Just open it, Cappie! Hurry! I can't stand surprises—mine or anyone else's." And she laughed like a little girl. I opened the box and peeked in; there was a brand new pair of ice skates! I was stunned. This was the nicest and most expensive gift I had ever received.

"Why, Inez, I don't even know how to ice skate," I said, picking up one of the skates. They looked big and bulky, but I was surprised at their lightness. Inez pulled another box from under the couch where we were sitting and showed me another pair exactly like the ones I was holding, except a bit smaller.

"I have been saving for something nice to celebrate being together for over a year now, and while we have been watching the skaters in

SHE WAS WIRED DIFFERENTLY

the park, I saw how much fun they were having and decided that is what we need to do, have some fun. I just finished the play. It is closing down for the winter months and they paid me so I used some of the money to buy us these skates." I couldn't speak; just seeing the happy look on her face made me happy. I kissed her right there in her parent's parlor.

Learning how to skate was great fun. We fell and laughed, fell again and laughed again, until we could glide around the pond with the other skaters, laughing, and hanging on to each other.

I got up early every morning and walked the short distant to the print shop. I was lucky because during the winter many small shops like ours closed down. The snow piled up on the streets so high that traffic was often stopped. But it was an exciting time. The city had over one and a half million people. The streets in certain areas were very crowded with horse-drawn wagons, horse-drawn cable cars, bicycles, lots of people walking, new electric streetcars, light carriages, and new automobiles. Traffic jams were everywhere, especially in the heavily populated areas such as the Loop. I liked the brisk walk, besides saving money. Sometimes Inez would walk part way with me. We would throw snowballs at each other and then hold hands and snuggle close to warm up. Sometimes we took one of the streetcars to Jackson Park on a Sunday. People accepted us as two friends now.

I was sorry for the poor working horses that I saw pulling wagons on the streets. They did not have an easy life. They were often imported from farms toward the end of their lives and worked the same sixty-hour weeks as their masters did. At night it was common for them, I learned, to return to the same crowded tenement districts as their owners, where they would occupy crowded stables on the rear of the lots, or even in basements. I thought of our ponies out on the Indian reservations, eating the prairie grass and sleeping under the stars or in a little lean-to shelter against a hill during the winter, with a blanket thrown over them.

During the long, hot summer months, we rented bicycles. The high-wheeled style had been replaced by a more manageable low wheel

HATTIE

and it became very popular. We soon liked bicycling better than any way to travel. It seemed there just wasn't enough time to visit all the places and see and do all the wonderful things there were to enjoy in the city. We bought ourselves two shiny bikes, put baskets on the handlebars, packed our lunch, and would peddle off for the day. Days like these became years; soon, five years had passed.

Inez and I went to a few church services on Sunday mornings; this was helping me put Mama to rest. I really wanted to understand what the purpose of life and death was. Mama believed in God and she loved going to Church on Sundays. I felt close to her in a church.

Inez was Catholic, so we went there first. I did not get anything out of that service because it was all spoken in Latin, which I didn't understand. Inez didn't understand it either because her parents stopped going when they came to America. The church was big and cold and I had tried to keep up with the kneeling, standing, and sitting by watching what everyone was doing. Inez started to giggle as she saw me watch everyone to anticipate the next move. I finally just sat down and did nothing, with a big frown on my face. After a few Sundays, she asked me, "Should we go to a church where they speak English?" "Oh, yes," I answered, "but you don't need to come if you wouldn't feel right about it." She answered me with a question: "Do you want me to come? If it bothers you that I am with you, I will not come!" I said "No, no, I love your company; we have so little time to spend together, let's not waste a moment." And I meant every word; I squeezed her hand and said, 'Let's try to find a church like my Mama's.'"

So we began to try different churches. Chicago was full of them. Some were big and beautiful, and some were small, like the little one in Salem that Papa had built. Each seemed to have part of the message that I was hoping to hear, but it always took me two or three services to begin to understand. I was looking for answers to some age-old questions: Where did I come from? What am I doing here? Where am I going? And I had some new questions for God, such as, Why am I so different?

The closest I came to appreciating the spiritual was what I heard

SHE WAS WIRED DIFFERENTLY

from the Sioux Indians about the Great Spirit that they believed shaped our lives. Because Indians link everything so closely to nature, and because living on the prairie drew me close to nature, this made more sense to me. Inez and I had many deep discussions about different viewpoints; I understood the reverence the Indians showed for the power of nature better than what I learned in church.

I did believe in the Ten Commandments and the Golden Rule. But I also understood the Indians' reverence for nature, the power of the wind, the gratitude for plants and trees, and the acceptance of life exactly the way things were. I knew it wasn't *our* way, but I saw the Indians not as inferiors but as people we could learn from.

I told Inez about an old Indian medicine man from a small Sioux tribe. Fred and I had met him out on the prairie one sunny bright day; he had his tipi under a small tree. His squaw woman was cooking something in a pot over a small fire in front of the tipi. We had seen the smoke and rode over to see who was camping there. We often would find a couple of Sioux men back from a hunt with their kills; they were usually on the way to the agency to exchange their kills for food or other things they needed. Being a couple of curious teenaged "boys," we liked looking at the animals and weapons the Indians used.

I remember this occasion well, for it was the beginning of a friendship with this old medicine man. He was very spiritual and had a deep awareness of nature, which he shared with us in broken English and many hand signs. We learned that the Indians had a way of looking at life very different from the white man.

"Death was greatly respected in all Indian culture because it is inevitable," I told Inez as she listened intently, looking at me with her deep blue eyes. "It was not feared or seen as the end of life; rather, it was regarded as a natural part of life, a time of transition into another world. Most Indian tribes believed that at death, the soul continued into an afterlife. I learned this concept from the old medicine man."

I talked to her for a long while about what I had learned from this proud, yet simple man. They believed in a Great Spirit. This Great Spirit has spiritual powers that influences the whole life of man and

HATTIE

every living thing under the sun. Our religions call it God and the Sioux call it "Wakan Tanka"; their meaning is "The Great Mystery." The shaman, as the old medicine man called himself, would draw us pictures in the dirt with a sharp stick of the sun, the winds, earth or a rock. All, he said, were Wakan Tanka, the Great Spirit. Everything is in circles, he said. "Life is a circle, everything circle, moon, earth, wind, no beginning no ending." Then he would pound the stick on the ground and said fiercely, "Wakan Tanka, the Great Spirit! No beginning, no ending!"

The U.S. authorities attempted to ban Native American religious rituals including the Ghost Dance, the Sun Dance, and the peyote cult. I told Inez about the last war dance that Fred and I watched just before our government forbade it. Of course, they had them after that anyway, as at the Rose Bud reservation in the Black Hills of South Dakota, where the Sioux had that huge uprising just before we left the agency for good.

"Papa thought it was all right for us to try to indoctrinate the Indians to Christianity, but to demand they learn and believe the Christian religion he felt was wrong," I said. "Mama and he had a few arguments about this. Mama usually won the argument, but I always felt that Papa was right. Of course, I didn't let on my feelings to Mama."

Fred and I came to believe some of the Indians' ideas about Wakan Tanka and about the mysterious forces of nature that supported human life. We also came to respect the shaman's belief that to survive as individuals, we must acknowledge and try to understand these spiritual powers in every aspect of our lives. Fred and I could see these ideas at work when we went riding and exploring in the wilderness near Chamberlain. We watched the seasons change, experienced the magic of the herbs the Indian women used, and saw the way the Indians used the animals, plants, and weather to provide for themselves. Listening to the shamans' and old men's stories, we realized that these practices had gone on for centuries. How could these things not be true?

"The Indians addressed these powers in prayers and song and dance, just like the Christians do in the churches we attend," I said to

SHE WAS WIRED DIFFERENTLY

Inez, "except the dancing and singing was very different." I laughed, telling her of the big dirt field and the howling of the Indians as they did their slow, up-and-down gyration while in a trance.

The hunting season, the trout runs in spring, and the appearance of sap in the maple trees were important occasions for the Indians. It meant there would be plenty of game, fish, and vegetation that year, so they gave thanks through dancing and celebrating. We witnessed many of these rituals.

I admired many things about the Indians' attitude toward nature. They never killed more than they needed. If they took something from the earth, they put something back, sometimes a seed or a small plant. They always honored the gifts of the Great Spirit, and gave thanks.

Inez was fascinated by the things I had learned. We talked and talked about what we believed and what puzzled us. She too had lots of knowledge about life from the literature and plays she had read.

Meanwhile my job was going along very well. I really liked it. I was learning so much about the printing of books, and thought for a while that maybe I should just stay in this business, as I saw Mr. Hatch was making a very good living.

CHAPTER TEN

"Chicago railroad terminals handled seven million long-distance passengers in 1903," said the headline in the Chicago Daily News. "I knew the city was busy, but I had no idea that it was that busy," I said to Inez. "That's a lot of people coming and going. We are in the center of the world and I wouldn't want to be anywhere else on earth. I love it here, it's so exciting!"

Inez was reading the funnies, her favorite part of the Sunday paper. She said, "I feel the same, Cappie. I feel so alive here."

Whenever we stepped outside, it seemed to be full of people walking everywhere, mostly on wooden sidewalks. A few old cobblestone streets were still around, but both were being replaced, we noted, by a new material called "concrete."

Inez and I walked everywhere or rode our bikes, mainly because it felt like we were alone even on the busy streets, dodging pushcarts and horse carts, gawking at the few new cars. We sometimes would jump on an electric car. It seemed as though we were always going someplace.

These were fun days for us in our first years. We would start our lazy Sunday mornings in the lovely parlor, reading the newspaper and having coffee. That was the only word I could use to describe her parents' large Victorian home—"lovely." It was set back on a tiny lot very close to the Loop. The location was worth a lot of money. Apartment houses had built up around it, and I was sure Inez's parents received many offers to buy the house for the location alone. They had inherited it from an aunt of Inez's mother. "Have your parents ever thought of selling and moving into one of the new areas?" I had asked Inez, coming home from one of our adventures. She had then explained how they came to own this lovely old house.

The mother's aunt was in poor health and begged for Inez's parents to come to America. Inez was only ten years old at the time; Aunt Julia needed someone to take care of her. She was in her eighties when they came. She lived another five years before she passed. "She was a bat," Inez said.

I looked at her quizzically. "What do you mean, 'she was a bat?'" I questioned.

"She was a patchwork of personalities, some good and some absolutely impossible to tolerate or to understand."

I had never seen this look of confusion and bewilderment on her face before.

Inez began explaining this aunt at length to me. "I learned how to act some roles in plays because of her. One minute she would be one person—generous, caring and kind—and almost in her next breath she would dress me down for forgetting some slight promise I had made to her. She did this in the most cruel and contemptuous way possible, usually in front of someone, leaving me choking with tears, especially when I was very young."

"Why was she so cruel?"

"I think she enjoyed humiliating people. It seemed to give her power," Inez said thoughtfully.

I could always remember the parlor room in detail. A wooden, carved velvet chair was facing a blue and ivory satin sofa. The polished baby-grand piano faced the front windows where all their friends gathered to sing their favorite songs. Every piece of furniture was exquisitely made and all the materials were perfect, from the drapes to the heavy thick carpet. It was like a stage setting. I described the home to

SHE WAS WIRED DIFFERENTLY

Papa when I wrote to him, but never felt I could do it justice.

Mr. and Mrs. DuBue had been on stage for years and still did bit parts now and then. Their voices could be heard all over the house and the neighborhood. They talked back and forth to one another continuously, bantering both in French and English. Sometimes it went on almost non-stop. I'm sure this fast-talking back-and-forth would disturb some people, but to me it was a happy sound.

Inez frequently apologized for her parents' loud, continuing discussion, but I told her, "Inez, when I lived in Mitchel I rented a room from a family who never said a word to one another. I never knew if they were home or not. Many times I would come into a room and they would not even say hello, unless I said something to them first." Inez looked at me in astonishment,

"Oh, Cappie, how sad, how did you stand it?"

"You are so sweet," I said, putting my arm around her tiny waist, "But I put up with it to get here, so I could meet you, of course!"

She looked a little puzzled, cocked her head a little to the side, looked at my smiling face, and I think gave up trying to figure out what I meant. She hit me with the newspaper in her hand, and said, "Cappie, you are so deep; you didn't know you were going to meet me!" We both laughed. We laughed at everything those days.

The DuBue home was warm and friendly. When I first met them, I called her parents Mr. or Mrs. DuBue when I needed to address either of them. Mrs. DuBue soon said to me, "Look here, Hattie, from now on you must call me Beatrice and Mr. DuBue wants you to call him Arthur. Now if you can't do that you can't live here." She laughed at my look of concern, hugged me, and said, "We want you to feel like part of this family." They lived a kind of "savoir-faire" attitude. They used that phase a lot. I learned it was French and meant: "The ability to know exactly what to do and the right or most charming way to do it." I wrote this phrase to Papa in one of my letters, saying that this is the way Beatrice and Arthur treated everyone who came to their home. They were charming! They had many friends, mainly show people who were all very social. Most of their friends had many talents—singing,

HATTIE

playing instruments, dancing. I was learning so much from them: It was opening a whole new world to me.

The house seemed always filled with people. The doors were always open. Inez said she could only take so much of the racket, as she called it, and needed to get away from it sometimes, so we took off on our bicycles during the summer and during the winter we would go skating.

I remember telling Papa how exciting Jackson Park was. A ten-piece orchestra played music every Tuesday and Friday evening. We read where the city was setting aside four acres near Washington Park to build an ice-skating rink. It was to have refreshment stands and warming cabins, and, most wonderfully, the rinks would have electric lights! I had read all about this park. Inez and I couldn't wait for it to open. I wished Papa could see me skate. I wrote him I was getting pretty good.

My job was good, and I was learning so much. Mr. Hatch had shown me everything he knew about the printing of books. I could do the work now without his supervision. I was happy.

Inez loved the arts. There were many museums in Chicago. We went through them slowly. She explained the paintings and the statues, the names and the artists to me. Often we went to see vaudeville and musical performances. Ticket cost ten cents for balcony seats. The seats on the main floor, front and center were thirty cents, which we didn't do very often.

The days flew by, then the years. Wonderful new inventions came along like the airplane. Mainly the new automobiles had everyone's attention. The death of Susan B. Anthony in March of 1906 broke our hearts. Inez and I had joined the movement and went to meetings whenever we could.

Mrs. Anthony left the fight for women's emancipation to her longtime colleague, Elizabeth Cady Stanton. Their lives were very different. Susan was a spinster who had devoted her life to the women's movement. Elizabeth was married and had children. Together they had established local and national women's organizations. They wrote books and pamphlets to support women's suffrage. They were planning

SHE WAS WIRED DIFFERENTLY

the first suffrage parade to be in Chicago in 1909.

I wrote Papa that I thought, in spite of Susan's death, the suffrage movement would continue. I asked Papa to write about it in his newspaper.

"It's only fair that women should be able to vote too, Papa. I know you must think it is fair now too, don't you?" And I waited for his answer.

Papa wrote me back, "Yes, my dear, I do believe women should get the vote. If anything, I think maybe 'they' ought to have a reading test for everyone before anyone can vote. Seems very few folks know anything about the history or what this country stands for."

I knew I could count on Papa to encourage us women. His views were not held by the majority of newspaper men in any part of the country. Papa always had a spunky attitude. It seemed he didn't care what readers thought. He could make his view on things sound right and fair, stirring up sentiment and interest without stepping on too many toes. Of course, there were times when he went a little too far. I think his drinking lent to his loose tongue at times, losing some folk's business. Some got so mad they quit the paper. But "controversy" was also what made him a good newspaper man and, as he said—laughing—"Lucky me, I have the ink!"

Then I had to write to Papa about Mr. Hatch closing the shop. He had been having trouble breathing. Walking up and down the stairs had become hard for him these last few years, and it was getting worse and worse. He needed to stop and rest every few steps. His darling wife, Karen, had passed about the same time as Susan B. Anthony in 1906. After this, Mr. Hatch seemed to lose the will to work and would come downstairs, look around, and leave.

He had been leaving most of the decisions to me, even to what to charge for our work. I was taking care of everything, so I really wasn't surprised when he told me he was closing up the shop. To my great surprise and excitement, he said he would give me a good recommendation with a friend of his at the Chicago Daily News. My dream of working on this great, big-city paper may come true, I thought.

HATTIE

It was late 1908 and I was thirty-two but passing for twenty-eight, with Fred's birth date. I went to see Mr. Hatch's friend. He looked at the letter and then looked at me over his large desk in the very big, busy newspaper room in the building I had passed so many times, wishing I was working in it. Six years had gone by. I didn't regret the time I had spent working for Mr. Hatch. In fact, it had given me the opportunity to slowly become the young man that I appeared to be. I was accepted as a man everywhere I went.

Mr. Carlson was a tall, lanky man wearing a green visor pulled down low on his forehead, shading his eyes. No one had ever seen him without it, I learned. He peered at me and said, "If Mr. Hatch says you're good, then you're good! We need someone to set type on the night shift. Pay is seven dollars a day, ten hours, six days. You want it?" This was more than I had been making. I was delighted and quickly answered, "Yes, sir!" I shook his hand and gave him a big smile. He told me who to see and where to go.

I ran all the way back home shouting the news to Inez before I even saw her. We danced and laughed and held hands and hugged; I couldn't remember ever being this happy. Beatrice and Arthur came running into the parlor and yelled in French, "Cappie, Inez, what has happened?" Inez told them, "Cappie has finally gotten the job at the big newspaper that he has always wanted! Isn't this wonderful?"

"And Cappie," she screamed, "I got the lead part in a wonderful play. Today is going to be our special day!" She grabbed my shoulders and looked at me intently. "Cappie," she said, smiling, "Can we make this a really special day? Can we tell everyone we are engaged?"

Her parents hearing this news just beamed at us. They put their arms around us and said to me, "We love you Cappie. Our Inez has never been so happy." I just stared at them, tears in my eyes, my heart so full of love at their acceptance of me, of us, of our relationship, that I just couldn't talk. We had stopped our dancing around. I had my arm around her waist and looked down at her first, then at them. I pledged to them, "I will take care of her forever." And I meant it from the bottom of my heart.

SHE WAS WIRED DIFFERENTLY

I learned my job on the paper quickly. I had so much experience by then that I didn't need much direction. My main concern was the fellows I was working with, a group of about six. They were pals with each other. Some were married but a few like me were not. They liked to go have a drink together, play pool, and talk about baseball.

I knew I just would not fit in with them, and I was worried that they might somehow detect my true identity. So I shared that I was engaged to this little actress, which made them understand that my spare time was taken up with her. They were kind of in awe of me knowing so much about the paper business. They also understood that I had some pull getting the job without going from the ground up, as most of them had to do.

Also, I felt they thought I was kind of an odd guy and out of their group anyway, going around with an actress who was playing a big role on the stage. They felt I was kind of a strange duck. I heard one of them say one day, "You're a strange duck, Cappie." And now everyone called me "the strange duck." I laughed with them and referred to myself as "the Strange Duck." This attitude seemed to help me out, so I laughed at myself and it made me one of them. Only they didn't know what a strange duck I was, I thought. All the practice I had being a boy with Fred had helped me prepare now for the role of my life.

One day, I had to hurry back from watching Inez in a matinee, so I hailed a cab to get to work on time. I was glad to be sharing the cost with another man, who had rushed for the cab at the same time I did. He was a bit older than I was. He introduced himself as "Mel." He was obviously very interested in me. At first I thought he may be bisexual and was interested in me for that reason, but after we exchanged names and a few pleasantries, he said very carefully and quietly so that the driver could not hear him, "Nice meeting you, Cappie Fuller. I hope you will not think I'm rude or out of line, but I need to be frank with you and I know we have only a very little time together."

He continued to speak quietly. "I know you are a woman cross-dressing. I know this, Cappie, because of your posture and your body movements. They are too feminine. Study yourself in a full-length mir-

HATTIE

ror and watch how men sit and walk. You are pretty good, but just a little too delicate, a little too hesitant. Men act as if they own the world, and their space." I was astounded and could feel my face getting hot. I could not believe what he was saying or what I was hearing, and by now I was sure he could hear my heart pounding in my chest.

How had he pegged me so fast, I asked myself. I felt like flying out of the cab but then I became very curious and asked, "How did you know so quickly? I have been passing for several years now and no one seems to have detected me. You are the first one!" I whispered looking at the back of the driver cautiously.

Mel quietly said, "I'm a woman myself, and have been working as a man for many, many years now. I'm an actor's teacher; I've been teaching people how to act on stage for many years. No one knows I'm a woman unless I want to reveal it to them. But I can tell immediately when someone is acting unless they are just about perfect. Of course even I have been fooled occasionally, but I can usually tell right away." I was astounded; I could not believe that this perfectly turned-out man who looked like he was well-off was a woman like me.

He quietly went on speaking while sitting back comfortably in the seat. "You will be able to recognize another cross-over, too, Cappie. In fact, I'll bet you felt something about me was strangely familiar, am I right?"

"Yes, I thought you may be interested in me. I felt myself blushing. Sorry," I said.

He laughed. "Most folks don't understand how very sensitive cross-over people are, but now you will be even more aware."

As we were almost at my building I said, "Is there a chance you would let me visit you sometime? I truly need to know someone like you and get some advice. I have been very much alone and lonely."

Mel pulled out a little business card from his pocket and shoved it into my hand, saying, "Cappie Fuller, I shall not forget you. Please do call on me." He said this with the warmest, kindest smile on his face. I almost wanted to stay in the cab and throw myself into his arms and pour out my heart to him. For the first time I had found someone who

SHE WAS WIRED DIFFERENTLY

knew me, what I was feeling, someone who had already experienced what I was going through. Instead I jumped out, paid my bill, and watched the cab drive off with my first true friend in it.

My eyes filled with tears as my heart was so full of gratitude. I couldn't wait to see this Mel again, and my next thought was, "I have something wonderful and exciting to share with my Inez tonight: an encounter with what I knew was going to be a very important person in our lives."

CHAPTER ELEVEN

I had been on the job at the Chicago Daily News about six months when I met Mel in the cab. I looked at his card dozens of times. It had an address that I knew was in the heart of the theater district near the Columbia Theater on North Clark Street.

There was a phone number on Mel's card. Inez's parents had a phone, but it was a party line and I knew the party sharing the line was nosey. Every time we would get a call, we could hear a click on the phone line and knew "they" were listening to our conversation. So I decided to drop by Mel's office. I thought about it for a week before deciding what to do. I wanted this first meeting to be just between the two of us. I hoped by a surprise visit I would catch him alone. I was very nervous.

So, on a day when Inez was busy, I hopped on a trolley and was soon standing in front of a two-story office building. Mel's street number was on a glass door. Looking through, I saw a staircase. I entered and noticed a plaque with a list of names. I found Mel's name and office number, directing me to the second floor. As I started up the stairs, I saw him coming down.

What luck, I said to myself as I saw that he was alone. I stopped on the staircase and putting out my hand said, "What luck, Mel, I was just coming to see you. Do you remember me? Cappie Fuller, we met in a cab."

HATTIE

"Of course, I remember you Cappie. So glad to see you, I've been wondering when you would show up. What perfect timing, I was just going out for a bite to eat, care to join me?"

I smiled, and said "Yes, that will be great." He grabbed my hand with his gloved hand.

Outside, he said, "There is a nice little café about a block from here. We'll go there Cappie, if you don't mind." I was delighted, first that I had found him so easily and also that he had time to talk to me.

"Perfect," I answered, and we both started walking. He was a bit taller than me, and a little heavier. His clothes were beautifully tailored. He looks as though he has money, I thought.

We turned into a doorway about a block away; I followed him into the small café. Blue booths were lined against the left wall and a long counter was down the middle. A cook was behind the counter at a big black stove, cooking in view of the customers seated at the counter. He had on tall white chef's hat and a white apron pinned with a big silver safety pin around his black-and-white checked cook pants. He yelled at Mel as we entered, "Hi Mel, beef stew today!"

Mel looked at me as we slid into the last booth in the row, and said, "Hungry?" A very young waitress came quickly to our table before we could say anything to one another. She sat down two glasses of water and two menus. Mel picked up his water right away and took a long swallow, shoving the menu back to her. "Beef stew and coffee for me." He looked at me and said, "This is the best food in town, but it's a secret, so don't tell anyone or we won't be able to get a seat in this place." He laughed; I liked his sense of humor. "Lunch on me, Cappie."

I blushed and said, "No, I'll pay, thanks," and to the waitress I said, "Make it two of the stews and I'll have coffee also." She left and I finally had his attention.

Mel sat back in the corner of the seat and looked at me with a warm smile and a curious expression, as if to say, "OK, Cappie, what is it you want to know?" I leaned on the table, my hands clasped in front of me, and started with the question that I had wanted to ask since I left him in the cab.

"What did I do to give myself away to you so quickly?" I spoke in a low voice. Even though the booths had high backs between them, I didn't want my voice to carry to any inquisitive ears.

Mel said, "Several things. First how old are you?"

"Thirty-two, but I only claim twenty-eight because I use my brother's birth date. Luckily, I don't look my age, or so my girlfriend tells me."

"Oh, so you have a girlfriend. Good. I'll want to hear about her, and maybe I'll get to meet her one of these days, but now I want to explain some very basic facts to you." The waitress interrupted again as she brought our coffee and fussed with the cream and sugar and silverware. I sat waiting for what seemed like a lifetime to hear what he had to say.

"Cappie, your clothes are all wrong. They were all right when you were in your teens, the little boy britches, the sweaters and a jacket, but now you need clothes that are tailored by someone who can fit a nice suit and pants to your shape and fill it out in the shoulders and across the back. The pants also need to be tailored to disguise your girlish shape better. I will take you to my tailor who can do this for you. You should wear a scarf as much as possible to hide the fact that you have no Adam's apple, and gloves to cover your delicate hands. But most of all, dear friend, I will show you how to sit, stand and walk like a man," he said with confidence.

I knew he was right, because I hadn't given any of these things much thought at all. I must be pretty dumb about living this life, I thought. Thank God Mel had the courage to speak to me in that cab. I knew I had found a valuable friend.

The beef stew came; it smelled delicious. I hadn't realized I was hungry. We both stopped talking and ate with gusto. "No dessert for me," Mel told the waitress, "But a little more coffee, dear." He winked. When she left, he said to me "It doesn't hurt your image to flirt a little with the ladies, but explain it to your girlfriend," he said with a smile.

The next thing he did was pull out two cigars from his inside coat pocket and handed me one.

I told him I didn't smoke and smiled inside, remembering the last time I had said that and the trouble it had gotten me into with Half-Moon many years ago. Mel said, "Learn; it will help your image, and everything we do to help our image helps us pass successfully."

I left him in front of his office. My head was swimming, partly from the cigar, from which I had just had a few puffs, and partly from all the information I was trying to remember. I stumped out the cigar, putting the butt in my pocket to practice on later.

On my way home I tried to remember all Mel's instructions to me and the attitude he was trying to convey. It made sense, the things I had never paid attention to—such as no Adam's apple, my hands, the style of clothes. I saw now how these little points could give clues to my true identity if someone were to get suspicious. They were vital if I wanted to continue without detection.

We made a date to meet in his office, so he could show me how to sit and walk; even the way I crossed my legs was important. He told me that he belonged to a group of cross-dressers. Most had been friends for many years, meeting once a month to exchange information, like where to find doctors, tailors, and barbers. "If you want to meet them, I'll take you to the next meeting."

I thanked him, grateful for the invitation, saying, "Yes, I would really like to meet others like me." I asked him what I could do to repay him.

With a smile on his face, he said, "Cappie, we are now brothers. We need to help each other in order to continue to live our lives with dignity and safety." I was touched. Someday perhaps I can do something for him, I thought.

I went home and shared all this information with Inez, including the cigar. She hated the smell as I did, but she said, "Cappie, I know what I'll do. I will learn to smoke cigars too, and then we will both stink!" I laughed, thinking she was joking, but she took the cigar from my fingers and puffed and coughed and puffed some more.

I met with Mel weekly, first at his office, then with his tailor and

SHE WAS WIRED DIFFERENTLY

his barber. He was strict and made me go over and over the walk and other elements that he thought I needed to practice. I become very self-conscious, looking at myself in mirrors and in store windows as I walked.

I had assumed that if I dressed and called myself a man, people would not question me. I knew I felt safer when I was with Inez because the contrast between us was so great; she was so feminine. It seemed almost natural to act mannish around her.

I met "the group," as Mel called them. There were three other "men" besides Mel and me. Also there were two ladies who were girlfriends of two of the "guys."

The meeting was held in a small bar, which had room for a few tables toward the back. We were seated at the largest table in the dimly lit room. It was late afternoon before cocktail hour, so we had the place more or less to ourselves. A piano softly played near the bar, a few people were talking to the bartender, and it seemed the perfect place for a group like ours to meet.

The first person I met was Marty. He was very elegant. Immediately, I felt he was wealthy. His clothes were perfect to my eyes. He was wearing a light brown suit with a very soft, lighter tan shirt and a dark brown tie. The outfit looked like something from a Paris designer. He was very soft spoken and had a casual, relaxed attitude.

I found out that Marty was from a family of royalty and had married a man while very young to convince the family that there was nothing wrong with her. Divorced shortly after the marriage, "her" parents provided her with a large sum of money each month to stay here in the States. Marty laughed when he told this to me; "They just don't want me to embarrass them," he said. He had gone to the best schools in Europe and could have held a number of top jobs, but he preferred to do nothing.

Next around the table was John. "He" stood up, shook my hand and with a great smile said, "I'm John, Cappie, nice meeting you." Later he shared that he had taken the name John from his ex-husband whom he had divorced shortly after they had married. He said, "You

HATTIE

will find this is a pattern with many of us. We seem to try marriage first, either to please our families or hoping it would change us. My ex-husband was a drinker and an abuser, whose only ambition was to get drunk."

Next to him sat an enchanting, lovely looking young lady. After John had finished introducing himself, he turned to her, put his arm around her shoulders, and said, "And Cappie, this is my friend, Nancy. She is the love of my life, so don't try to steal her from me." He laughed, but I sensed he meant it.

I took Nancy's slim white little hand and said, "Nice meeting you, Nancy. Don't worry John, even though Nancy is pretty enough for me to want to steal her away from you, I have a love of my own. Her name is Inez and next time we meet, I'll try to bring her along. Right now she is in a play." I had to throw that in about Inez because it made me proud to have an artist like her in my life, and I guess I needed at this moment to brag to these people because I so wanted to be accepted.

Joe was the last of the three "men." He was wealthy, bright, witty, and handsome. He would not have fit our stereotype if someone just looked at him, for one could not tell whether he was male or female. His passion was art. He painted, with much success, I learned. With his success as a painter came power. He, too, wore expensive men's clothes, but clothes that could be worn by either sex at this point in time. Clinging to his arm was a beautiful, blonde young lady. I learned Joe used her as his model and bragged that she could pose for hours without moving a muscle.

Joe and I became friends immediately. I loved arty people, like Inez and her parents and their friends. It seemed to me they were more open and accepting of life just as it was.

Joe was flamboyant and eccentric, I learned. He had a large studio where he held lavish parties after finishing a painting. But while he was painting, no one saw him. Sometimes he would disappear for months and then reappear, present his newest work, have a great party, keep everyone laughing with his stinging wit until he seemed to get his fill of people, and withdraw again.

SHE WAS WIRED DIFFERENTLY

When I met with Mel's group again I brought Inez. She fit right in. The girlfriends had seen her on stage and knew she had the lead in the play at the Columbia Theater. They gave her great compliments, which made her feel very welcomed and special.

We had a sandwich and a beer with them, talked a little, and planned our next get-together. It was to be at Mel's place, and everyone said, "You will love his place. He is a great host."

We began hanging out with "the group."

We invited Mel to our house, feeling he would know many of the people Beatrice and Arthur knew in his profession as an actor's teacher. And he did. He had been brought up in show business. His mother had been a fairly famous actress and his father had been a producer of many plays. Both his parents were still active in show business, although his mother had retired from the stage several years before. His father financed a play now and then.

Inez and I had decided to get married after our engagement when I had gotten my job on the newspaper. We talked about how we could do this with the least amount of trouble. We shared our marriage plans with Mel and asked if he knew of anyone who would perform the ceremony for us.

He knew of a pastor who had a lesbian daughter. She had committed suicide several years before. This pastor had vowed that he would make his life a memorial to his love of her, so he preached and spoke of tolerant love to a small group of supporters. Sadly, he could not find a church where he could preach since he spoke out so frankly about Jesus' love for everyone including, he felt, people like his daughter.

It had been a year since our engagement. We had made a commitment to our relationship, and to each other. We wrote out a pledge that we wanted to share with Inez's parents and some friends of the last several years.

We contacted the pastor and went to see him. We found him a wonderful, warm, friendly man who willingly sat and talked to us about marriage. He had lots of advice and information. He felt any two people who were vowing before God and the world of their love for one another must understand what those vows meant. So he went over them, carefully. When he came to the children of the union, he said, "Sadly I guess you will not have this experience, but if for some reason children would come into your lives would this be all right with you?" We both laughed and said, "Father, I don't think that will happen, but if it did it will be all right with us."

The wedding was in the parlor of Inez's parent's home. Extra chairs were placed in every corner, but still some friends had to stand against the wall for the short ceremony. Inez wore a beautiful white satin gown, a pink sash around her tiny waist. She carried a small bouquet of baby pink roses. I wore a beautifully tailored black suit, with a white starched shirt and a pearl white tie.

We knew we looked elegant. Mel verified this fact. The first one in the line of well-wishers, he beamed his approval, shaking my hand he said, "Elegant, elegant, you both look wonderful!"

"High praise from the perfectionist that he is," I said to Inez afterward.

We had our picture taken. I sent one to Papa, Fred, and each of my sisters. It's about time, I thought, that they all know who I am now. I was proud of Inez and our relationship.

We had no time to go on a honeymoon, as we both had our jobs. It was July and hot. We chartered an electric boat for two dollars an hour at Washington Park, which took us all around Lake Michigan. It was fun. We went to dinner at our favorite restaurant, Berghoff, on the Loop.

The very next summer of 1911, Inez's parents passed away a few months apart. They had contracted the flu and it had attacked them both almost at the same time. Inez and I nursed them, along with several friends of theirs, but sadly they didn't survive.

Beatrice passed first; she was only fifty-five. Arthur, seeing her pass, did not even fight when the flu struck him. He was ten years older than Bea, as he always called her. He just went to bed and seemed to go to sleep shortly after we buried Beatrice. Inez had not even had time to grieve for her mother when her father passed.

SHE WAS WIRED DIFFERENTLY

It was like losing my own parents. We cried as we went through the house, remembering them in all the dear things they used and loved—their clothes, a shaving mug, the garden that Bea toiled in. We had a lovely service in the same parlor where we had gotten married the year before. All our friends and theirs came. The little pastor, Mr. White, who had performed our wedding vows, helped us say goodbye to these two most wonderful and kindest people that I ever knew. Inez and I inherited the big old house and we continued to live there.

Papa and I agreed that this war in Europe, which had been talked about all over the country for a year by now, was not any of our business. We argued with everybody about going in, but I was beginning to feel lots of people wanted us involved.

Papa's paper, The Western Call in Reeder, North Dakota, was just north of the state line between North and South Dakota. Fred's horse ranch was fairly close to Papa's paper, near Buffalo in South Dakota.

Fred also was clerking at a bank in Buffalo and was running for county treasurer. He had gone into politics and was trying to get the town of Buffalo declared as the county seat. Hearing all this news over the past few years, I understood Fred not wanting to be associated with me in case I was found out. That would look bad for him, especially if he and I were having a social relationship. It made me feel sad; we had such a strong bond together while growing up, and I missed him.

I shared all these feelings with Inez. She really didn't understand his fears even though I had tried to explain that the Dakota people were very different than city folks, and way different than show people (especially show people from France)! But she said, "Cappie, your Papa doesn't seem to have those attitudes, and he is from the Dakotas. Why is this?"

"Well my dear, Papa was never one to care much what people think and as long as he is running the only paper in town, no one can print anything against him." I laughed. "My papa has always been fearless, I think because he lived among the Sioux Indians. His family, the Fullers, are descendants of the Pilgrims, who came over on the Mayflower, so he has come by fearlessness honestly, I guess."

HATTIE

I think Inez had a better understanding than I did of the significance of that historical event. As an educated person, she was interested in history and art, and knew all about the Mayflower. She was pleased, I found, to learn that my family had such an important role in the country's history.

"Well, Mr. Fuller," Inez said, "Whenever I get a chance to meet your Papa I will be quizzing him about his Mayflower connection."

Papa sent word that my sister Mabel was the editor of another paper, The Dickinson Recorder, in North Dakota. He sent me a picture of her that appeared in the paper. Mabel was featured as editor. I knew then that she must be well-known in that area. Pretty good for a girl, I thought. I wondered what happened to Emory. Papa didn't seem to know.

Nellie was still married to Dan Carey and still living in Salem, South Dakota; they had a pack of kids, six I knew to date. She had a couple of sons who would be about the right age for this war in Europe if we were going to get in it.

The war clouds gathered and Europe went to war in 1914. As 1914 turned into 1915, all the men at work were making plans with each other that if the country got into it they would all go in together.

Papa wrote me when he heard that news. He thought it was time for me to come to North Dakota. He would fix me up, he said, with a job in Hettinger, with a small town newspaper, called Adams County Record. He felt I would be safer than in Chicago, in case of an army draft. Meanwhile, he said we could look for a paper to buy and a place for us to live near him in North Dakota.

I talked with Inez about this. She was in her thirties and parts on the stage were becoming scarce. When she saw my dilemma, it scared her. "What if they have a draft," she said, "Oh Cappie, we must sell and get you to a place where you are safe!"

We sold the house and moved to North Dakota just before America entered the war. I felt confident passing as a man now. I had been doing it for so many years, and of course having Inez on my arm was a big help.

Mel warned me as he was saying goodbye, "Cappie, you have cap-

SHE WAS WIRED DIFFERENTLY

tured the male identity. I watch you now and I know no one could ever detect your secret, but there are still places and situations that we both need to be careful of, like accidents, doctors, and men's bathrooms. We have learned how to avoid these situations. We are pretty safe, but never forget the act you are playing, my friend."

CHAPTER TWELVE

Papa found me a job right away at the Adams County Record in Hettinger, North Dakota. It was a short distance from Reeder, where his paper, The Western Call, was located. Both of these small towns were near the North and South Dakota state line, not far from the Montana state line.

I loved the newspaper business; I was good at the technical part. But I saw I needed to bone up on my reporting skills. I needed people skills. Inez was better than I was at meeting and talking to strangers. She quickly learned all about the people in Hettinger—their names, what they did for a living, and who was married to whom. She and Papa talked about the best way to get information and they decided that we should join everything—every club, social circle, and especially church. So, that is exactly what we did.

That first year in Hettinger we were busy just getting adjusted to the weather. Chicago had cold winters but North Dakota was a lot colder and for a lot longer. People native to this area didn't let the cold weather stop them very often from working or getting around.

One old fellow said to me, "Young fellow, if we closed up shop every time we had a bit of snow, we wouldn't get anything done. You'll learn to work right through a snowstorm." I saw he was right. People here were very tough, working in snow up to their hips.

We learned that if you didn't plan carefully, you could die if you were caught out in a snowstorm without the proper preparation.

When spring came, we began to scout around for a town that had a newspaper for sale. Papa wrote, telling us about a paper in a small town called Rhame. It was northeast of Reeder. Inez and I took a train one day, stopping in Reeder. We picked up Papa, and the three of us traveled on to the little town of Rhame.

We liked the town right away. We walked from the train station to the newspaper office. We looked over the equipment, and checked the circulation, which was very small. We saw at once that they didn't have much to sell. The press was old, and the paper was apparently not being read much by outlying farmers. But Papa thought it had potential for growth.

I figured that I could offer to print business flyers and personal announcements, but the one idea that excited me the most was to set up a printing company for books.

The paper was up for sale cheap. On January 1, 1917, we took over. Three months later in April, America entered the war in Europe. Inez and I found a small little house and we began to meet the people of Rhame.

Papa had set up a printing company called "The Fuller Printing Co., Publishers" and he brought us into that business. In North Dakota, Mabel was the editor of the Belfield Review in Stark County. Having Papa and Mabel in the newspaper business gave me the legitimization I needed. Even though I had been living as a man since 1904, thirteen years now, Chicago was quite a different place than a small town where everyone knew everybody's business

Gossip seemed to be the only entertainment these folks enjoyed. I understood that from my experience living in Chamberlain and Salem. In Chicago, people were busy with their own lives and more or less minded their own business. But here, I knew, we would need to be extra careful. We needed all the cover we could get, so we started out very cautiously and kept our mouths shut.

Inez found it hard to understand people in a small town. She had

SHE WAS WIRED DIFFERENTLY

always worked with open-minded people. It was hard for her to be on guard constantly about our views. The townspeople seemed to want so many details about us. Inez knew how important this was for our safety. "Don't worry, Cappie, I shall pretend I am on stage playing a part. I can do it!" And so she did, capturing everyone's heart while answering many questions.

I was very careful at first. I slowly got myself involved in civic and school affairs. Inez began helping with school plays and gradually got acquainted by volunteering for various community-service clubs. Our life was pleasant and we communicated frequently with Papa. But I was sad at the loss of my brother and sister's friendships.

Nellie's two older sons, Edwin and Marc, had gone into the service and were fighting the war in Europe. I was glad we had decided to move when we did; the war was raging and every day more and more men were being sent over. The trains coming through town were full of young men in uniform leaving from our area.

Women were going to the train station bringing food for the men as the trains rolled through. The two churches in town organized groups to put together packages of personal necessities and little gifts to lift the spirits of the boys going overseas. The women also gathered to roll bandages to be shipped to the medical personnel at the front. Everyone was involved in some way. Many of the young men in our town were talking about signing up.

I was concerned that people would wonder if a young man like me, without children, should be signing up. Although I was thirty-eight, I was still using Fred's birth certificate and it made me thirty-three, a reasonable age to join, many would think. It was unsettling because I didn't want the government to scrutinize my records too carefully if they started a draft. Finally, the war ended November 11, 1918, and Inez and I, along with the whole town, turned out for the biggest celebration it had ever seen. It was a cold day and a big bonfire was built in front of the city hall where everyone went to celebrate.

The next thing we knew, the 1920 census was being taken. We listed ourselves as Alfred D. Fuller and wife Inez. I found out that

HATTIE

Inez had spoken to Papa about how much it grieved me not to have any contact with my sisters and brother. She thought it seemed un-Christian, and asked, "Don't you think so, Papa?"

Papa got busy and went to visit Nellie and Fred by train. I found this out later. It was a long trip for him, especially down to Salem, South Dakota, where Nellie and Dan Carey still lived. Papa was sixty-seven years old. He was still a strong man; and he seemed to enjoy visiting the family.

I don't know what he said to Nellie and Fred, but after his trip, he came to us for a visit. He said to me, "Cappie, I want you and I to go visit Nellie and her family. It would do you good to make the trip. Nellie lost her youngest little boy, Melroy, you know he died last year of scarlet fever. Nellie has been feeling poorly every since, I think a visit from the two of us would cheer her up a little, what do you say?"

Inez encouraged me. "Do go Cappie, I'll stay and take care of things. I've learned to do what needs to be done on the paper for a short time." I held her very close that night and told her "You are the best thing that has ever happened to me Inez, how did I ever find you? How come I deserve someone as special as you are? I love you with all my heart."

Papa and I left on the train for Salem. I had not seen Nellie or Salem for over twenty years, and found it had not changed much. Dan, Nellie's husband, met us at the train station. He was still working for the Great Northwestern Railroad and was a section foreman now, a strong Irishman with a ready smile and a twinkle in his eyes. Edwin, their eldest son, was there, with his dad, to help us with our bags. He was a big strapping young fellow who had graduated from college as an engineer just before the war, and then had gone into the army. He had been a young boy when I left Salem. His younger brother Marc had served in the navy during the war and, thankfully, they both came back safely.

Besides the two older sons, Nellie had another son, Eugene; two daughters, Ethyl and Nellie; and then two more sons, Clarence and Melroy. Melroy, her youngest, had died. The boys, of course, didn't

SHE WAS WIRED DIFFERENTLY

know me. I could tell that Nellie had not told any of her children about my secret and I was grateful for that. Papa had told me on the way that none of my sisters or Fred ever discussed me to anyone. Dan and Emory, of course, knew, as did Nellie, Fred's wife, but they all kept my secret. They did this for their own protection as well as mine. I was a shameful topic.

Papa told me Nellie was afraid her kids would ask questions, because they had seen pictures of our family and they knew she had only one brother, Fred, whom they had met. "So Papa," she asked, "what can I call *him*?"

Papa advised her that, if pressed by one of the children, she was to say, "Go ask Papa."

"What will you say?" Nellie asked anxiously.

"I'll give them some hog-wash about a brother of mine." He laughed. So that's how we handled it and it went smooth as silk.

It worked so well that I even had my picture taken with everybody. My youngest sister, Ethel, was also in town visiting Nellie with her new husband, Chris Rygh. They were on their way to California. We had a nice visit and talked about everyone wanting to go to California.

Papa and I stayed at the Diamond Hotel for two nights. It was the same hotel where we lived for a time twenty years ago, when Mama had quit the boarding house business. The hotel hadn't changed one bit. It was like I had stepped back in time, I said to Papa. He laughed saying, "Nothing much changes in these little Dakota towns, Cappie."

Nellie and I were able to have a few brief talks alone during my visit. I told her how sorry I was for any trouble my change had caused her. She asked no questions and seemed extremely uncomfortable talking about my change in appearance, avoiding eye contact. Her only comment was, "Well, I always thought that you and Fred riding out on the prairie among the Indians was not a very proper life for a young girl. I talked to Mama about you several times, the way you wore Fred's trousers and acted just like a boy. I didn't think that was good for you." She was silent for a few moments; we were sitting out by her garden.

I tried to explain my feelings. "I choose to act like a boy, even then.

HATTIE

Don't you see how I was even then, Nellie?"

"Well, I suppose so," Nellie replied. I could tell the conversation was over. She got up and made an excuse to go inside. It was obvious she did not want to discuss it with me any further.

The evening before we left, I spoke to her about her little five-year-old son. We were again seated outside, alone in her little garden.

"I am so sorry about Melroy, Nellie."

"Cappie," she said, feeling at last comfortable with the name everyone called me now, "no one can know the pain of losing that little fellow. He was such a delight. I have taken up nursing babies to get out of the house, because every where I look I see him,"

Shortly her head was on my shoulder and she was sobbing. I patted her shoulder telling her how very sorry I was. "Nellie, there must have been a reason for his brief life. Try to figure it out, for the God of my understanding does not make mistakes. There is a purpose and a reason for everything under the sun." She clung to me, listening. We were sitting outside in a swing overlooking her vegetable garden that reminded me so much of Mama's gardens all those many years ago.

"Oh Cappie, I hope you are right. I will try to remember that and look for the reason for his short life."

"When Mama died, Nellie, I too experienced pain. She was so dear to me and I felt my news had killed her. Maybe you and the others thought so too?" I said this while still holding her hands in mine. I tried to look in her blue eyes, much like Papa's, to read her response before she spoke.

I felt Nellie looking at me with compassion. She said, "No, Cappie, I never blamed you. I knew Mama had been sick for years off and on. You never knew because she hid it well with medicine, but I think they gave her some wrong medicine the last time and so does Ethel."

I couldn't believe my ears. Here I had been carrying around this guilt all these years and knew nothing about Mama's illness. Of course, Nellie would know this, working so close to her for so long.

We sat in silence for a long time. I released her hands and looked up at the full moon above our heads in the dark sky. Then I began talking to her as if to myself, like I do when writing in my diary or talking to Inez.

"Nellie, I have asked myself over the years, why am I like I am? Did God do this to me as a punishment, making me different than other people? Where is this God everyone talks about? Why must life be so hard, especially for women? Why must people suffer so much? Like Mama and Papa had to do to just scratch out a living for themselves and us children? I have gone to many churches looking for answers and began to love the parables that Jesus spoke, like the story of the Good Samaritan.

"I remembered when I listened to a wise old Indian medicine man when we were on the reservation. He taught Fred and me to look at what the Great Spirit has given us, the sun, moon, rain and all the living things, even the peyote which when smoked shows us another dimension to life. He said everything has a soul, and he said we are only here for a moment. He didn't say moment; he just smacked his hands together, pointed to the sky, and said, 'Light in sky. Here!' Then smacked his hands again and said, 'Gone! Be ready.' Pounding his chest he had said, 'Be brave, be proud, take care of earth, trees, water, children, the old, the sick.'

"I have come to this conclusion, Nellie: We are here for a moment to find love for ourselves and to give love to others on our journey. Each of us has a mission; for some it will be a long mission, for others a short one. Our purpose is to bring joy and to find joy, but it is also to learn the lessons for our soul's purpose. Some of these lessons are for our existence and some are for others to learn from us. The burden that we learn will help others to carry their burdens."

We then sat for a long time, saying nothing, both of us in our own thoughts. Finally Nellie said, "Thank you Cappie, I want you to know, you are welcome here anytime. I can tell you are happy and that makes me happy. I too have been given much. Thank you for talking to me about your beliefs. It has helped me very much."

I felt that night, while lying in the hotel bed, that perhaps this was one of the reasons I had come here to visit my sister, to share with her

HATTIE

my perspective of life. If I was able to help her, I was glad. When we had parted for the last time that night, she said, "Cappie, I'm so happy you came," and she gave my hand a tight squeeze.

As I lay my weary head on the train's nice, soft lounge chair on the way home, I said to Papa, "Thank you for clearing the way for this visit, I'm very glad I came"

Papa patted me on the knee and said, "It went well, didn't it?"

He treats me like a son now rather than a daughter, I thought to myself. Funny, I had always been more like a son to him. We had shared the same love of the newspaper business and had worked together like a father and son.

Papa said, "Cappie, I have written to Fred about your paper and your Inez. I told him what a nice person she is, and how the town of Rhame has taken the two of you to their heart. Fred wrote me back that he and his wife Nellie are planning on coming to visit you." I was stunned, because I knew in my heart that Fred was worried about me being discovered and what that would do to his reputation. This was a great concession for him and for me.

"I can't tell you Papa, how much I appreciate your support over all these years, and now you have helped me connect with Nellie and her family. If I can connect with Fred and his family my heart will be full. You are the best father anyone could ever have!" I so wanted to cry. But that is one thing that Mel always said, never cry, so I had trained myself to think of something else right away when I felt vulnerable. Sometimes I would pinch myself hard on an arm or leg, which seemed to help me take my mind off of any sentimental feelings.

I said, "I know it has not been easy for you, having a child like me Papa, but I want you to know I love you, so much, especially for your acceptance!" He simply looked at me with those blue eyes, a bit faded now, but still full of love for me. He patted me again on the knee saying, "You are still my little chick-a-dee, and I love you too."

CHAPTER THIRTEEN

PROHIBITION
WOMEN GET THE VOTE
Prohibition: Amendment 18 ratified January 16, 1919,
effective January 16, 1920
Women Vote: Amendment 19 passed Congress June 4, 1919,
ratified August 18, 1920

Inez and I were elated that the House of Representatives passed the 19th Amendment by a vote of 304 to 90 and the Senate approved it 56 to 25. It still needed to be ratified by the states. Illinois, Wisconsin and Michigan were the first to ratify.

We were proud that Illinois was the first; we *just knew* it was because "our" Chicago groups of women were so active. It was still amazing to us that it had ever been in dispute that women couldn't vote and that it had taken an amendment to the constitution to get our fair rights.

We were now full citizens under law. Of course, we knew it would take many years for some men to accept a woman running for office, let alone be elected president, but we believed that someday this could and would happen.

Nellie was very happy working for the Women's Christian Temperance Union, which was the most influential advocate of pro-

HATTIE

hibition. They had worked as hard for Prohibition as the Susan B. Anthony groups had for the women's vote.

I had made no bones about my opinion on women being able to vote and found women sneaking up to me at church, away from their husbands, whispering, "Thank you, Cappie, for sticking up for us," and then scurry off.

We knew there were still many men who did not believe that women should vote. Some men said loud and clear that they would keep "their" women from going to the polls, but we felt all women would sooner or later demand their rights. The passage of the amendment was a big occasion for us.

I was thinking about my life as I watched Inez moving around the kitchen, placing the napkins, cups, sugar bowl, and creamer on the table as if she were performing on stage.

We both at times had become bored during the winter being so cooped up in our little home; in response, we had found writing projects. Inez was working on a play she had started writing while in Chicago. I was writing different stories about growing up.

I remember writing about Papa taking Fred on different occasions, just the two of them. Papa would say, "Fred, get ready, I want you to go with me to get our supplies today."

I remembered the day I asked, "Papa can I come with you?"

Papa had looked at me kind of puzzled and said, "Hattie, you're a girl, you need to learn how to help your mother. Fred needs to learn to do men things." And that was that. I never asked again, but I was hurt and asked myself what Fred could do that I couldn't do?

I sometimes would question Fred about what they did and what they said while getting the supplies. Fred didn't understand my interest or what I wanted to know, and I don't know if I knew what I wanted to know, either. But it just intrigued me, that man's world, a world I was shut out of.

Papa taught all of us the newspaper business, thank God, I thought now. I remembered going with him to his newspaper office as we were growing up. I always went; Fred and Mabel would go too because they loved the paper business. Nellie never came nor showed any interest. She was Mama's biggest helper. She liked to cook and care for the younger girls.

Papa showed us, with so much patience, everything about laying out the pages of the paper and how to insert the pictures on a page. To me, it was fun. I loved the work. It was like a big puzzle, and I enjoyed figuring it out. This brought me back to Inez again. Reading and writing was something that drew us together. We both seemed to be looking for the answers to the eternal question of why we were so different from other people.

During winter, while the snow piled up outside, and the logs burned in our fireplace, we read and shared our thoughts about ideas. We enjoyed each other's company. I felt lucky to have such a wonderful partner.

As I sat in the kitchen, all these thoughts seemed to fly through my head, chasing out my anxiety about Fred and Nellie's coming visit. I felt at times as though I couldn't wait to see him. Then I would get this sinking feeling in the pit of my stomach, wondering how they really felt about me. Did they feel disgusted and think of me as a freak? How would they treat Inez, I asked myself a dozen times.

"Maybe I shouldn't have tried to contact him," I said again.

Inez answered, "Cappie, you will never know how he feels until you look into his eyes."

She was so wise, I thought, because I knew she was right. I would know exactly how Fred felt about me immediately, for I did know him well. I told myself that if he were just coming to please Papa and didn't want me in his life, I would finally know.

Back and forth my head went as the time for their visit drew closer. I remembered when I last saw him, as he was packing up to leave for the service. He seemed so young in 1901, and of course, he was young! Twenty then, almost forty-two now, and I was forty-six. Papa was sixty-nine and next year he would be seventy, an old man by everyone's standard.

I wondered how I had let this much time go by. We had been living in Rhame for six years now, not too far from his ranch in South Dakota. These years had just flown by. He was a father and I had never met his wife, Nellie, nor seen his children.

"Where had the time gone?" I asked Inez.

She just looked at me with a knowing gaze and said, "I ask myself that all the time, Cappie. My parents have been gone for over ten years now, yet it seems like yesterday. I talk to them every day, asking them this and that question, and they answer me." She came over to where I was sitting, sat in my lap, and put her head on my shoulder. "I still miss them, Cappie."

I hugged her to me and said, "I know, my darling, I miss them too. They were such wonderful parents to the both of us."

Finally the day came. Fred and his wife Nellie got off the train with Papa at noon. Our home was close to my newspaper office and fairly close to the small train station, so we had asked Papa to walk them to our house.

The townspeople liked us and we were invited everywhere. We were considered "Pillars of the Church," mainly because we never missed the service unless we were snowed in, or sick. I was on the school board as well as a member of the businessmen's club in town. For this reason, I didn't want to meet Fred for the first time at the station. I knew if anyone saw us, they would of course be curious and would want to be introduced to these out of town visitors and this would be awkward.

"Here they come," announced Inez, excitedly pulling the curtains aside and looking out our front window. She had been peeking out since we heard the train whistle. Our home was lovely, I thought. We had a nice big fireplace in the living room, crisp white lace curtains in our windows, new electric lights, and indoor plumbing. There was a large, screened-in front porch across the whole front of the house. We sat out there during the hot summer months catching the breeze and eating dinner at the wicker table on many warm evenings.

When Inez said they were coming, I knew I would hear them climb the stairs first and walk across the creaky porch. I waited listening intensely. I thought my heart would jump out of my chest.

SHE WAS WIRED DIFFERENTLY

I finally heard their steps crossing the porch. I saw Papa coming through the door followed by Fred. We both just stared at one another. He was somewhat dumbfounded I'm sure, for I looked nothing like the girl he remembered when he left on the train those many years ago. I could tell he didn't recognize me, but to me he still looked like the Fred I knew. I could have picked him out in a crowd.

He was fuller of face and body and had a few wrinkles, but he looked about the same to me. We always looked somewhat alike, with dark brown hair and eyes. He was a little taller than I was. He had always had a kind of direct look, which kept most people at a distance. I knew that I too had this effect on people. Inez said I had an educated manner and attitude, a kind of no-nonsense air about me. I could see what she meant when I looked at my brother after not seeing him for twenty years.

His eyes and attitude said, "What's this all about Hattie?" Although he didn't put those thoughts into words, I read his puzzled expression, and knew I would have to convince him of my soul's desire.

Growing up, he had seen the difference between our sisters and me clearly at times. When I dressed like a boy, wearing his old trousers and putting my hair up into that old cap, he knew I was different.

All these thoughts flashed through my mind quickly as he introduced his wife, Nellie, to us. I introduced Inez. I didn't say she was my wife, for I didn't want to shock them anymore. I knew seeing me in men's clothes with a man's haircut was enough change for now. I introduced her saying "And this is my Inez."

Inez had planned a lovely dinner of roasted chicken with all the trimmings. She said, "We are so thankful for your visit. This is a Thanksgiving dinner." She had outdone herself. Everything was perfect.

I asked about their children while we were eating. "I have seen many pictures of your children," I said. "Papa is so proud, he seems to have a new photo every time we see him." I laughed. We had a nice conversation about their three girls and their son, Fred Jr.

Fred said, "I see Papa keeps us all up to date on family." I could tell that Fred had began to relax when he patted Papa on his shoulder and said "Yeah, he really gets around. He goes to see all of us kids. No one would know he is almost seventy!"

"Yes," I agreed, "Just look at him!" We all looked at him proudly and he seemed very pleased with himself.

"Well," Papa said, "We have the newspaper business kinda tied up in this part of the Dakotas, with Mabel up North in Stark County, as an editor, Cappie here and me in Reeder."

I noticed that Nellie had gone into the kitchen to help Inez with the dessert and I could hear them talking. I even heard a little laughter, which made me feel a whole lot better. When they came out to serve Inez's apple pie, I was able to exchange looks with her. She smiled at me so I could tell that all was going well. I finally really began to relax.

Papa had begun talking about the paper business, his favorite subject. Fred and I listened. He always had the latest news.

All papers ran humorous drawings, a great delight to readers. Papa said that the comics section was the first one that people turned to these days. Everyone wants a good laugh, but the sports page is still the most popular. Men and boys loved to read about games. Men loved reading the inning-by-inning, play-by-play descriptions of all the ball games. "You know young people learn to read while following their favorite sport. The big papers also report now on horse racing, bicycle racing, rowing races and any athletic events these days." I had put some of these events into a special entertainment section in my own paper and was watching to see if it increased my circulation.

Fred and Nellie had planned to leave the next day. After supper, I took Fred and Papa down to my newspaper office and showed them the lovely little printing press I had picked up cheap from an older publishing family that was closing out their book-printing business. I loved this part of the printing business and was recently receiving orders from all over the Dakotas and beyond to print books. My reputation had spread for the printing of beautiful books, I told them with a proud smile.

I especially loved to print books of poetry. I showed Fred and Papa

SHE WAS WIRED DIFFERENTLY

some of my best work. I knew Fred was impressed by the questions he was asking. He shared his dream of owning a newspaper of his own one day soon. Papa was observing us with a big grin on his proud face, as we talked his trade talk.

The next day, the three of them left on the train. Inez and I were finally alone and after putting the house back into order, we sat down with a cup of tea, and talked.

Inez said, "Well, Cappie, we sure worried a lot unnecessarily, didn't we? They seemed a little surprised, but you know, it didn't seem to take them long to accept us did it?" I needed to be reassured. "Did you really think that Nellie accepted us, I mean me, as she and you seemed to get along well."

"I felt that Fred accepted me as Cappie fairly quickly, especially when we went to the paper and he saw the printing office. I think he was a bit in awe but I never connected with Nellie. I did hear the two of you talking in the kitchen. I even heard you laughing. How did she seem to you?"

Inez came over to the chair I was sitting in and put her arm around my shoulder. "You know Cappie, we are so lucky. First, my folks liked and accepted our love for each other, then your Papa then your sister Nellie, and now your brother and his wife."

I pulled her down onto my lap and said, "Tell me more." And smiled wanting more assurance.

"Cappie, they felt comfortable enough to travel all this way to visit us. I think they did this out of love for your Papa, and Fred's memories of you. But, I felt that Nellie really liked us and was glad they had come. You know what she said to me after you all went to the office?"

"No, what?"

Nellie said, "I'm so glad Fred finally agreed to come and visit you and Cappie. He had talked so often about his sister Hattie I mean his pal, Cappie. Then she and I both laughed."

"Nellie told me she had hoped that Fred would accept Papa's invitation to come and visit with him.

"And then she said the sweetest thing, Cappie. She said, I'm so glad

HATTIE

he did, because I got to meet you, Inez. Wasn't that kind of her?"

I was speechless, not believing my good fortune. This was about the best news I could have imagined after their first visit. I looked at her and let the tears flow down my checks. To hell with Mel, I thought. Just this once, I'm going to let myself go and feel this feeling of gratitude.

CHAPTER FOURTEEN

The roaring twenties! Most people believed Woodrow Wilson, laid low by a stroke and distracted by the war, had mishandled almost every problem presented to him. Papa and I editorialized in our papers about what people were saying about Mrs. Wilson running the country.

Most folks didn't seem to pay much attention to the League of Nations, or to Europe's problems, especially here in North Dakota. We had all retreated back into the shell of isolation.

Inez bemoaned the fact that she and her parents had been born too early to work in the movies. She also loved the radio, and had her ear glued to the speaker of our tiny radio for hours.

Prosperity was real, our paper was selling well, and my printing business had taken off. The town was doing great and everyone was getting all kinds of new-fangled equipment for their farms.

Inez and I began going to Fred's ranch in South Dakota on the train. Sometimes we would bring one of their children back with us. Papa would also often meet us either at the ranch or at our home, and we would have a family get-together on the holidays. This made me happy, but seemed to make Inez very emotional. She was captivated by the children and became excited, planning for days with special surprises and things for them to do. She eagerly waited for each time we would get to see them.

HATTIE

Fred's daughter Maye was twelve going on thirteen. She had become fascinated with Inez, who gave her pieces of jewelry and clothing that she loved. Also, because we had one of the first radios, she and Maye would listen for hours. Inez taught the girls how to dance the Charleston, which was the rage. She made clothes for them on her new Singer sewing machine.

We bought a "Tin Lizzy," as the new Fords were called. Both Inez and I learned to drive. People in town were amazed to see her behind the wheel because she was so little; she had to sit on two pillows to see out the windshield, but they were not surprised seeing us engage in trying all the newest fads. We were different because we had no children and had come from the big city. They knew Inez had been on stage, so the women were in awe of her. The idea that we had no children was heartache for Inez.

"Cappie, don't you miss having a child?" she asked.

"No, my dear, I love my life the way it is."

At first I was glad that Inez had such an interest in my brother's children because I knew that this was the one thing she had always missed in our relationship. I had seen this longing in her many times. She loved to take care of the children in the Sunday school classes. She would watch them while their parents attended the pastor's message. The parents in Rhame really appreciated her for this, and many would say, "It's too bad, Inez, that you and Cappie haven't had children. You would make such a wonderful mother."

I had found her crying sometimes and asked her why. She always said, "I love you Cappie, but I do wish I had a little child sometimes. There is just something inside me that wants to be a mother to a child."

I didn't know what to do to make up for the loss in her life. When we were younger, in Chicago, I sometimes felt guilty and became cold and withdrawn from her. I knew I could not give her what she wanted. She felt me retreating from her and would ask, "Why are you being so cold to me, Cappie?"

"I wish sometimes you would leave me." I answered.

Then she would plead with me, "Why, why?" Yelling at me in her most theatrical voice.

I would say in despair, "Because, my dear, I can't nor never can give you a child, which you want so much."

Then she would put her arms around me and kiss me saying "Please, Cappie, don't ever leave me. You know we are soul mates."

I promised never to leave her.

Still she would question me again, "Cappie do you wish for a child?"

"No," I would say quickly and with as much conviction as I could, because I meant it, "I do not want nor need a child in my life." I wanted to make it clear that my life was complete with her. I didn't add, I wished that her life with me was good enough but I thought it.

This scene had played out in our lives over and over and though I wished each time would be the last time we would have this emotional exchange. I knew I would be going through it once again; it seemed she needed to express this sadness.

But as we got older, it seemed her need for a child had diminished somewhat. Since we had moved to Rhame and were involved with the church, she had found many children there whom she could mother. She held a little girls' dance class once a week during the summer. It seemed for a long time that these activities had satisfied her maternal needs.

I thought it was good that she had become so fond of Fred's children, but I soon saw it was becoming an obsession. She didn't seem to be able to think of anything else and she became especially attached to Maye, who I must admit was also completely enraptured by Inez. They hugged and kissed and everywhere Inez went, Maye would tag after her. Fred and Nellie were pleased that Inez seemed to care so much for their girls.

I had been having some chest pains. Through my old friend Mel I had found a doctor in Baker, Montana, when we first came to North Dakota, whom I had been seeing. Dr. Weber had a sister like us living as a man. Our network of friends in Chicago worked well, but they could not find a trusted doctor closer to me, so I traveled to Montana twice a year, telling my friends that I had family there.

My cover story was safe, as Dr. Weber did become like a family to Inez and me. Over the years, even Papa had accompanied me a few times and had met the gentle doctor and his family. I really hadn't had many physical problems in my early visits to him.

He seemed interested in my childhood and had asked many questions about how I had felt when I was growing up. One day he shared with me that his one sister also had desired to dress and acted like a boy while she was growing up. She became a doctor, which was a big accomplishment for a girl in those days. Dr. Weber was very proud of her. She was practicing in Chicago and was Mel's doctor.

Dr. Weber over the years told me about another sister, a writer, who was writing a book about gender malfunctions. She was writing case histories about her sister and also her daughter, who seem to have the same feelings growing up as her sister. She feels trapped in a girl's body.

The causes and the ability of parents to understand these children and help them feel good about themselves was the purpose of her book. The doctor said we all know so many who had committed suicide. They would become especially despondent if their families rejected them. The doctor wanted to interview Papa, as his attitude was so unusual.

Papa and the doctor had long conversations, because fathers seemed most opposed to a child crossing over. Fathers usually said to the child, "Just get over it!" And then they would try to make a boy child tougher, by trying to get them into football or fighting. A girl child they tried to make more feminine, by giving them dolls and insisting on dancing lessons. These fathers usually rejected the child, and turned cold, and these children were typically the ones who took suicide as a way out of their unhappiness.

Some of these children had been born with both male and female genitals. In these cases, the doctor asked the parents for permission and decided on one set of genitals in the first months after birth. Sometimes they were right but quite often they were wrong and then the child suffered for the rest of his or her life.

SHE WAS WIRED DIFFERENTLY

Dr. Weber had many long talks with Papa. He took my history down and asked me many questions. Papa and he sat for many hours talking. He came several times on his own, helping the doctor with the history of gender, from his viewpoint.

My chest pains had started about 1923 when I turned forty-seven. I began making more frequent trips to Montana. In addition to the medicine the doctor prescribed for me, he told me I needed to lower my stress level. I shared with him Inez's desire to have a child, and how crazy she was about Maye. Dr. Weber thought it was just a passing fancy, and recommended I let her indulge herself with playing mother to the girl.

Inez seemed to get more and more frustrated because she was unable to see Maye very often. She wrote to her and lavished presents on all the girls, but they could seldom come to visit. They had school and when the weather turned bad, weeks would go by without a word. I told Inez, "Just accept the time you can see them, my dear, or you will make yourself sick."

One day shortly after Thanksgiving, the snow seemed to be a mile high, but the weather was good enough to get to church. So, Inez had gone and come home from a women's gathering. She came in with some wonderful news, she said. Her prayers had been answered; her cheeks were bright pink and her blue eyes were sparkling as she gave me all the details.

"Guess what Cappie?" she asked breathlessly. "You know how much I've always prayed for a child! Well Guess what! I think my prayer has been answered, but not just for one child, but for two, two girls!" She was fairly shouting as she went on to tell me the whole story. A dreadful accident had occurred only a few days before. "I took down some notes for you Cappie, because I know you will need to write a story about it for the paper." As Inez went on telling me the story, I got a nervous, sinking feeling in the pit of my stomach.

The parents of the two girls had been killed in a freak snowstorm. It seemed a bridge had collapsed as they were driving over it. The girls, luckily, had not been with them. They had been left at church for

HATTIE

a children's choir, which was preparing for the Christmas pageant. The parents had planned on coming back for them later that evening. When they didn't return, a search party went looking for them and found their truck down in the creek of running water with both parents dead inside.

These people were poor farm workers on a farm owned by an old German couple. The mother was helping out in the farmhouse and the father was their field hand. The couple had come from the same small town in Germany as the farmers owners. Inez thought they were distant relatives. None of these people spoke much English, and the girls had only gone to school here for a short time.

The pastor and his wife were taking a personal interest in finding the girls a home, as the old German couples were not healthy enough to raise them. They would not be able to teach them English or help them get to school or church. So everyone in Rhame was looking for a family in town that could at least help out or, even better, to adopt the girls.

Inez breathlessly finished with, "Cappie, don't you think this is a sign from God that we should help these two unfortunate little girls?" She spoke on and on about these children, and the anxious feeling inside me got stronger.

I calmly said to her, "Sit down Inez, let's have a cup of tea and talk this over."

I was looking into her blue eyes. I did not want to hurt or deny her anything, as she was everything to me. It was going to be hard to dampen her enthusiasm about these two girls, I could tell. Inez already felt they were "ours," but I so needed her to understand my reluctance in bringing any child into our lives, so I went forward putting my feelings into words.

"My dear soulmate." I took her delicate, white, slim hand into mine from across the table. "I must speak to you about my fears. First, I remember my mother with all us children to take care of. Believe me my dear, it is really difficult and expensive to care for children. I am over fifty and you are not far behind. Look at Fred and Nellie. They

SHE WAS WIRED DIFFERENTLY

are younger than we are and Nellie tells us all the time how hard it is to care for their children doesn't she?"

I paused and Inez said, almost pleading, "Yes, Cappie, I can see that it takes a lot of energy, but I still have lots of energy, don't I?" She asked like a little girl.

I answered her truthfully, "Yes, dear heart, you do, you have much more than I do, but this is not all. I think you know what I am going to say. I am not what I seem. I have passed as a man quite successfully, so far. I'm even on the school board here in town, but those two little girls are going to grow up living very close and they surely will find out about me. What then? What then my dear? Is that really going to be fair to them?"

Inez grabbed my hands tightly and said, "Cappie, you are who you are. If they cannot see your soul as I can, then we will not have taught them anything!" I squeezed her hand and said, "I cannot deny you this experience of being a mother, my dear. I will do my best at being the best parent I can be, like Papa is." I gave in.

We got the two German girls. I was pleased that the younger one was named Elizabeth, reminding me of my mother. She was six and her sister Ellen was nine. They were very shy, frightened, and sad at the loss of both their mother and father in a strange land where they knew no one and didn't even speak the language. I knew no German but Inez with her French background seemed to be able to communicate almost from the first day. They watched and listened to us, and picked up a few words. Soon they were talking in sentences.

I would see Ellen holding Elizabeth, who cried often at first. She would rock her back and forth like a mother, speaking little German words into her ear. They were bright girls, and were grieving their terrible loss. Inez brought them from church for a "try-out," they were told.

It took us awhile to be able to have a conversation with them, but over the next few months, which the try-out turned out to be, they grew closer and closer to Inez. I pretty much kept a distance between us, for fear of them learning my secret. Inez felt this was wrong of me. She thought the girls experienced my lack of warmth as a rejection of

HATTIE

them. Discord entered our relationship for the first time.

We all loved to hear our beloved Lawrence Welk's orchestra; he was a native son of North Dakota, but was also from a German background. He spoke English with an accent, which seemed to endear him to the girls. Lawrence was born in 1903, and had a little group of musicians called the Hotsy-Totsy Boys. Everyone all over the Dakotas was dancing the Charleston to his music. Lawrence himself was only twenty-five. He had started the group when he was only thirteen years old. The girls would forget their loss at times, and learned to dance the Charleston with Inez. Her enthusiasm would catch on with them and their sadness seemed to dim over the few months they were with us.

Inez's attention seemed to be totally on the girls now—their clothes, their schoolwork, and their activities. She had them in her dance class and in all the church youth activities. She had to be home when they were due home from school, so she could give them a little snack and hear the reports about their day. Then there was the washing and cooking and cleaning to do. I did miss the companionship we had before the girls came but I was happy that Inez enjoyed her mother role. I began to busy myself more and more with the printing of books, which I loved.

But I missed Inez at work. She had always been at my side and I didn't realize how much she had helped me. But even when she was there now, her mind was not. She was totally involved with the girls. Was I jealous? Of course I was, but I tried not to let her know.

The roaring twenties gave a new definition to womanhood. A new woman was born even here in North Dakota. In the movies, women were smoking and drinking in public. They danced, worked, and lived independent lives, especially the younger generation. I liked their carefree attitude, short hair, and movie-star make-up. They showed confidence in a new lifestyle, and they seemed happy that the Victorian lifestyle was in the past. But Inez was not interested in this women's newfound freedom. She was totally immersed in her new career as a mother.

CHAPTER FIFTEEN

Headlines were screaming in October of 1929 that the booming stock market had crashed. This aggravated the fragile economy in Europe and soon became part of a worldwide depression. Inez and I were also having trouble. The country's problems started about 1928 and by 1929, everything had gone downhill fast. I was fifty-two years old.

Papa had sold his paper in 1924 to Harry Goodell, Nellie's nephew. Papa had kept right on working for Harry, showing him the ropes, as he called it, which meant teaching him the newspaper business. Finally, Papa decided Harry could run the paper on his own and he traveled out to the West Coast to see some relatives.

I was in a deep depression about Inez and our relationship. Inez had become very involved with trying to adopt Ellen and Elizabeth. Another couple wanted to adopt the girls. The girls had gone for a brief visit to their farm outside of town. This decision by the church to allow the girls to visit the young couple, who had shown an interest in them, broke Inez's heart, and she became frantic while they were gone.

Inez believed that the only reason the couple wanted the girls were for them to work on the farm as they got older. I tried to reason with her about the situation, saying, "You know Inez, it seems that a farm, like they lived on with their parents, would be more like home to them." She just glared at me and did not speak for several days.

HATTIE

I could tell that she was becoming more and more desperate. Then she began pressuring me to speak out about our home. "Can't you say, Cappie, that our home is the perfect place to raise the two girls? We are in town near schools and church. This location would give them a more enriched life than living way out on an isolated farm."

She pleaded her case to the pastor and his wife, as well as her women's group. I heard her say shrilly, "Look at all the advantages we could give the girls than living on a farm."

"After all," one lady friend said to me, "Inez is not seeing that this young couple is much like the girls' parents were, young and living on a farm. I think it would feel more like home to them. They could always visit Inez and you in town, Cappie. We must think of what is best for the girls."

I agreed with her in my heart, but did not say this to Inez.

The girls came back from the first visit, and as children can be, were brutality honest.

"Oh, Auntie Inez," they said (Inez had instructed them to call her Auntie and me Uncle). "We loved the visit to the farm, when can we go back again? There is a little lamb just like Mamma had. We both got to feed it," Ellen said. "Anna allowed us to give it the same name as Mamma's lamb. Isn't that good?"

On and on the girls talked, telling Inez about what they did during their visit. In their broken English, they said, "Anna said we could come back any time. Could we go again tomorrow? Please?" they pleaded.

Inez with tears in her eyes hugged them and said, "Yes, my little dears, you can go again, real soon."

My heart broke, but I did not know how to comfort her. Whatever I said just seemed to make her angry. She glared at me and turned away. It was as if I had turned into the enemy along with the pastor, his wife, and the women in the women's group. She had always enjoyed being with her women friends, but now avoided them all. She felt the whole town had turned against her.

Inez came to accept that our age was against us, but still believed that I was against her for my own selfish purposes. I could not convince her that I would have done anything to make her happy. Believing that everyone was against her, she finally gave up in her heart and agreed to let the girls spend more and more time on the farm as this seemed to be what they wanted.

Inez stopped going to the group and to church. Some of the women called at the house for her, but she barley talked to them.

The pastor came by after she missed church services several Sundays in a row. She would not come out of the bedroom even to greet him. I explained to him that she was still mourning the loss. He asked me if he could stop by again. I said, "I hope you will, as she is really heartbroken."

So he stopped by the very next afternoon. I was home and let him in. Inez put on a false smile and asked if he would like a cup of tea. He said yes. I said "I'll fix us all a cup." I retreated to the kitchen, hoping the pastor would get her to talk.

I heard him say, "We have all missed you at the church services, Inez. Can we talk about the reason you are not coming?"

I could barely hear Inez's answer. It sounded like "Everyone has let me down."

When I came back with the tea their conversation had only gone on for a short time and I felt a strained silence between them.

I said, "Pastor, it has been hard on us to have those two lovely little girls in our home, feeling we were a family for those months and then to lose them because someone else wanted them."

The pastor replied, "Yes, I can understand your feelings but I have been explaining to Inez, we need to do what is best for the girls. You and Inez are wonderful people and there is no reason you could not adopt a child. But many of us saw that the girls felt more at home on the farm, like the home they had been growing up on. So for that reason and that reason alone, we encouraged this solution. It seems we were right. The girls are very happy and that is what we all want for them, isn't it?"

I said, "Yes of course."

Inez echoed resentfully, "Yes, yes of course."

HATTIE

The pastor's visit went on for only a short time after that. I could tell Inez was still hurt. She finally said, "I understand that you all thought the farm was more suitable than *our home*, and that's fine." Her whole attitude was of one who was being rejected unfairly.

Soon he left, knowing I'm sure that he had not accomplished his mission to comfort her, as he had hoped. He could see what I had seen, that this went deeper than what rational thinking could fix.

We packed all the girls' clothes and everything Inez had purchased for them. I remembered the three of them pouring over the Sears catalog for many hours and then waiting patiently for the postman to bring the packages to our doorstep. What fun those days were for Inez to see the girl's pleasure with their choices and they gave Inez big hugs and kisses, which she had adored. As she helped them pack, I knew her heart was breaking and I could do nothing. I tried everything to comfort her; she acted as though I was a stranger and that my words and actions were not genuine.

After her initial shock, she vented her anger in different ways. She didn't cry or throw things; she just became very cold and withdrawn. One day she announced she was going to take a trip to Chicago to visit friends.

I tried to cheer her up and remind her of the many times that I had spoken out, in town and in church, regarding my feelings for the girls—about how sweet they were and that we loved having them join us. But Inez's only comment was, "You sounded fake and insincere."

She blamed me mainly because she knew in my heart I did not want to raise these girls—or any children—and she was right. But I had adjusted to having the girls around. I had told her this over and over again. I would have done anything for her. I had never forgotten the promise I had made to her parents—that I would take care of her forever. I knew in my heart that I had tried my best, but I knew also that my best had just not been good enough.

It had been a year of battles, so I was hoping that this trip would heal some of the wounds and help us get back to our old selves. I really thought she would be back soon and I think she thought so too. I had

SHE WAS WIRED DIFFERENTLY

no idea on that snowy winter day, December 13th, 1930, that when she got on that train I would never see her again.

I missed her so much, but I knew that it was the old Inez I was missing. At first I was relieved at her absence, as the last year had been the worse year of my life. At times, I had felt as if I never really knew Inez. She was cold and snapped back at me in such a disapproving tone of voice that I just stopped offering any advice. I finally didn't speak to her at all unless she asked me a question.

After she was gone I became very depressed. I did not care about anything. The country was also in a depression; I followed the state of the economy, which took my mind off of my troubles at least for a little while. My news sources were talking about the reasons for the depression. They were calling it the Great Depression because every country in the world was being hit by it.

Papa had seen this coming. He was wise enough to sell his paper to get out from under, and he had come and talked to me at that time about me selling too. But I was in the midst of our difficulties with the girls, so I couldn't think about selling out then. Papa said, "Cappie, this is just the beginning." He, of course, was so right.

My health was so bad that I had all I could do to get the paper out every week. I had to give up smoking the cigars and I had a hard time even walking to the office. It was only a few blocks, but it seemed I had turned old overnight. I didn't care whose fault the depression was and it was not like me to be disinterested in the news.

Fred and Nellie had seen some of my health problems as we had been visiting them on their ranch and they knew about the girls. They knew also of my reluctance to adopt them. Nellie seemed to understand Inez's feelings and had tried to explain them to me. And I tried to explain myself to her, but I don't think she could understand me either.

They knew after Inez left that I was sad and lonely and knew also that my health was not good. So they came to help me when I needed to go to Montana to see my doctor. Thank God, I thought, Fred knew the newspaper business as well as I did. I felt so lucky that they would do this for me.

HATTIE

I heard the good doctor say, "Well, Cappie, my advice is for you to take a vacation, a long one; get away from all the nervous tension." At the tiny smile on my face, he said, "I know, I know I should take my own advice, but believe me my friend you are either going to have a heart attack, boom! Like that! Or more likely a stroke and that's worse in a way depending on which part of your body shuts down."

He saw the despondent look on my face and added, "I'm not trying to scare you Cappie, believe me; this is not scare talk. But unless you change your life right now you won't live another six months. I can't put it any plainer to you because you must hear this. Your blood pressure is way up and that pain in your chest is your heart talking to you."

He took my charts and said, "I'll see you in my office and I'm giving you another prescription for your blood pressure." While getting dressed, I thought, how long had I been coming here? Was it 1912 or 1913 when I first came to him? I tried to remember. I noted how slowly my brain was working.

I didn't seem to care about anything. "Let's see, where was I?" I asked myself. Oh yes, how long have I known the doctor, since 1912, and its now 1931. That's almost twenty years! Seems longer, seems like I've known him my whole life.

I went into his office. He was behind the familiar old desk piled high with papers and books. He shuffled through the papers and finally came up with his prescription pad; he sat down in the dark, hard-backed chair, pulled it up to the desk, and began writing on the pad.

I avoided his eyes for as long as I could. Finally I looked at him and could tell by the serious look on his face that I was really in trouble. He sat back in the chair and said, "How old are you, Cappie, the truth?"

He opened my file to check my age. "Almost fifty-five," he said. "I'm telling you Cappie, and your ticker is telling you. Change your lifestyle, relax, have some fun." He peered at me with those steel grey eyes of his that seemed to miss nothing.

"Where's Inez?" he asked. "Is she back from Chicago yet?"

"Doc, I don't think she is coming back. I haven't heard from her. She was so disappointed about the girls. She felt the whole town had turned against her and that I too had turned against her." A sob caught in my throat. "I tried to tell her that this was the best thing for the girls, but she just couldn't get over the disappointment."

"I know your health problems sounds frightening, Cappie," he said, "and you have no idea how to carry out this prescription I just gave you about change. Everything has gone to hell, it seems, especially, I hear, for you folks in the Dakotas, that black blizzard we are hearing about, has taken all the crops?"

"Just about," I answered.

"Can you sell and follow your Papa out to the West Coast?" he asked me. "That sunshine and fresh air is just what you need, my friend."

I looked at him with no hope in my eyes but I tried to put on a happier face. I gave him a little smack on his white-covered arm as we stood together and I said, "Thanks again, Doc, you are such a great friend. I don't know what I would have done all these years without you. I will try to figure out how I can follow your directions. I'd sure like to do exactly what you said—go West and follow Papa. I'll see and I'll let you know." With that, we said goodbye. I left feeling for some reason a little hope. Just maybe, I thought, I could take a trip to San Francisco. I had always wanted to go there.

I got home late the next day and sat with Fred and Nellie at the kitchen table. I began explaining my doctor's visit to them. I was surprised at the smiles on their faces in place of what I thought would be worried looks.

Reaching across the table and taking my hand in hers, Nellie said, "Cappie, we have been talking since we have been here. The girls love it here. The town is so nice. They have made friends. Our ranch is so far out of town, and the girls have no friends there. Both of the older girls will be in high school this year." She took a deep breath. Letting go of my hand she said to Fred, "You tell him, Fred, what we decided to ask him to do."

"Well Cappie," Fred said, "we were thinking. And knowing your health has been pretty bad, it looks like you could stand a rest. So we

HATTIE

thought maybe you would like to sell the paper to us. Our ranch is not doing so well. Nobody has any money or need to buy horses anymore. I think we can run the paper and the book business, too. I've learned enough to keep it going." He looked at me to sense my reaction.

"I can't believe my good luck," I said, as this wonderful suggestion began to sink in. It seemed to be the solution to my prayers and my problem. I felt a smile spreading across my face, and to think I was going to be helping my brother and his family with their problems, as well as helping myself—what luck!

I said, "You two are a gift from Heaven to me at this moment. I can't believe my good luck. Yes! Yes! This is all I have been praying about on my way home. I want to go to San Francisco, and I can't leave fast enough." I laughed for the first time in months.

I moved out and they moved into the house. I knew Fred would take care of the paper. We made our financial arrangements, so all he had to do was send me a little money every month to keep me going. And I was off.

About this time, both South and North Dakota began suffering the worst drought in history. It was called the "black blizzard." Fred, Nellie, and their children moved into my home and ran the paper beginning in 1931. During this time, it was utter despair in the Midwest, all over the country, and in my life as well.

I worried that Nellie and Fred were going to have a hard time making ends meet, but the book business brought in some money. The house and business were paid for so they didn't have rent or a house payment. I knew they would be better off than on their ranch.

I was in San Francisco when Franklin D. Roosevelt was elected. I was a Republican, but I jumped my party line, like a lot of other Republicans, because times were so hard for everyone. We needed a change, I thought, and I liked Roosevelt's attitude, summed up in his statement, "There is nothing to fear but fear itself." I wanted to believe this not only for the country but for me too.

I had been feeling so depressed about my health and about Inez leaving me. I no longer paid much attention to the news the way I used to. I was living from day to day. I remember walking every day along the waterfront in San Francisco.

Why did I move here, I wondered. Maybe I should have gone to Chicago and tried to locate Inez. But the memories of our last year would return and I knew in my heart that it was over. We had a wonderful, loving relationship for a long time but it was finished. I knew that now. I felt like an old horse that had run the race but lost at the end. Still I prayed. I felt God hadn't brought me this far to drop me. There was still a mission for my life.

I had been writing to Papa regularly. He was living in Bakersfield, California, with my sister Elizabeth. I hoped that I could get to see him soon. I had kept him up to date on how bad Inez felt about losing the German girls. "Papa," I explained, "Inez just did not seem to get over the disappointment of losing those girls."

Papa wrote me back and said he felt sad for us, but he developed an insight about women: "They seem to need something to mother," he said. "It's just part of their nature." He always understood people, and he was in his eighties.

The next I heard was a note from Elizabeth telling of his passing:

Papa passed March 30, 1932. He was 81 years old. I am sending you a copy of his death certificate.

It was signed by Elizabeth. I cried all night; Papa was more than my father, he was my best friend.

CHAPTER SIXTEEN

For the second time in several days, I sat next to a guy with dirty fingernails at the counter of a small café on Market Street close to my hotel. I loved San Francisco, even though I had only been here for a couple of weeks. I was still looking the city over, trying to decide what to do.

We said "hello" as I sat down. He didn't seem like the "dirty fingernails" type. The rest of him was "clean as a pin," as Mama would have said. I noticed he had a paper under his arm. After he ordered, he took it out and began reading the front page. I couldn't help looking over at the headlines. I began evaluating the front page, and was impressed. The print was crisp; the lettering was of a good size for reading. So many papers were printing in small type these days to save money on ink and paper, but it made reading difficult. I realized I was in a big city that sold lots of papers, so they could afford larger type.

The young fellow noticed that I was looking at his paper with interest, and he turned toward me and with a charming smile said, "I can leave you this paper when I'm finished eating. I work for the newspaper and can get another one free." I said to myself, Bingo! That explains his dirty fingernails. I should have known! Shows me how distracted I've been. I couldn't even identify a fellow newspaperman.

"Thanks, I would really appreciate that." I smiled back. "Didn't mean to bother you," I continued, "But I was curious about the paper.

HATTIE

I'm fairly new to San Francisco and looking for a job. I'm a printer; I've worked on a few papers in my life. My name is Cappie." I put my hand out to him. "It's nice meeting someone who's in the newspaper business."

"I'm Matt", he said, and extended his hand to me. We shook. He said, "I think they have a position open for a printer. It's for nights, doesn't pay much, but it could get you a foot in the door."

I was very grateful. "What luck," I said, "I will take anything."

I looked him over more carefully. Matt was tall, blonde, and pale—not very attractive in my opinion, but of course I was not attracted to men, and he seemed more like a boy to me. He was young, had a great smile, kind of like a kid brother.

I asked him where the office was and told him again I really appreciated him telling me about this job. Jobs were very scarce, so this was a big favor, and I knew I was lucky to have met him. He said it was just down the street and if I was done eating, he'd show me where it was on his way back to work.

The economy all over America had gotten worse and worse. People were selling apples on the street for a penny and there were long lines of men waiting to apply for jobs. Hamburgers only cost seven cents and our coffee was a nickel, but many people didn't have even that much to spend. Newspapers sold for two cents.

We walked the block and a half to a big newspaper building and entered through a side door to an office. We went up to a small desk where a woman was sitting; she was a fairly tough-looking older lady, and looked to me like she had been there for years and was part of the fixtures. She smiled at Matt, and said, "Hi Kid, how's it going?" Matt answered, "Hi Mabel, I brought this friend of mine, Cappie, in to apply for that typesetting job. How about it, is it still open?" She looked me over and reached into a drawer. She handed me a form, saying, "Yeah it's open, just fill this out Cappie, and I'll see it gets to the right guy." And she gave me a wink and a smile. The wink seemed out of character, but I smiled back.

"Thanks, thanks a lot." I took the form and sat down on a chair

SHE WAS WIRED DIFFERENTLY

near a small table. Matt came and sat in the other chair, and I looked at his smiling face.

"Boy do I appreciate this Matt, how can I pay you back?"

Matt stood up, put out his hand to me with his winning smile and said, "I get off in two hours, if you want to buy me a cup of coffee. I'll meet you back at the restaurant. Good luck."

I shook his hand and said, "Great, thanks again, I'll be there."

I filled out the form, gave my background in the paper business, and handed it back to Mabel. "I have a sister named Mabel, so I won't forget your name. Should I come back tomorrow to see if I got the job?" I asked her. She took the form and looked it over.

"You just sit right back down there, Cappie. I'll take this into Carl right now and see what he has to say." I was so glad not to have to wait for an answer.

"Gosh Mabel I sure appreciate that," I said. She smiled at me and I felt like a fool for sounding like a kid, but she didn't seem to notice; she hurried off.

She was gone for it seemed like a long time. Her telephone rang twice, and then was silent. I sat back down in the chair and waited, saying my Mama's prayer again: "Please, help me Lord, and I'll do my best to help you."

Mabel came back with a big smile.

"You come back in the morning and see Carl. He can't see you right now, but said he would talk to you tomorrow morning at ten. Be here on time, Cappie; he is a stickler for promptness."

"Thanks very much Mabel, I sure appreciate your help." She just smiled a very nice smile. I changed my mind; she didn't look hard or tough at all. "I'll be here at ten sharp." And I left in a very happy mood, forgetting my troubles for the moment.

I went and sat in a small park for the couple of hours waiting for Matt to get off work. I read the newspaper he had given me and then I walked back to the restaurant. I had only been in town for a couple of weeks. I had walked around, rode the little cable cars, explored Chinatown and Fisherman's Wharf. What a city! I thought. No won-

HATTIE

der everyone said it has everything. It did! I was beginning to feel a lot better. I was like a little kid, staring at everything with astonishment. I had lived in the big city of Chicago, but that had been many years ago, in a very different time. This was some city, I thought, and wished Inez was here with me, the old Inez.

I went back to the restaurant and sat at a little table for two. I ordered a cup of coffee. Matt came in a few minutes later and ordered his cup of Java. I looked him over carefully. He had a friendly and easy manner and as we talked, I could tell from his vocabulary and the subjects that he was fairly well educated. I felt that this was the beginning of an enjoyable friendship.

The next morning I got the job. It started that night, from 12 a.m. until 9 a.m. Lousy hours, lousy pay—only fifty cents an hour—but people were dying on the streets from lack of food. I saw desperate people every day, so I was very glad to have work. Now I could write to Fred and tell him not to send me any money. I wouldn't need it. I knew this would take a big load off his shoulders. It made me feel good not to have to depend on him. I knew he would be having it tough with his big family to take care of.

The newspaper business in 1932 was booming. Many people had only the few cents it cost to buy a paper. It was about the only entertainment most families could afford. People shared their papers with friends and neighbors. The help-wanted section was well-read. Shops and restaurants only had to put a sign in their windows for dishwashing jobs or for busboy jobs and they were filled in a few hours.

So I felt very fortunate that I had a trade that was in demand. Famine was being reported throughout the world.

When Roosevelt received the Democratic nomination for president and began campaigning it seemed to bring a new spirit to the country. Many people experienced a little hope when Roosevelt defeated Hoover by a wide margin. Right away he began unfolding his New Deal program.

Matt and I had struck up a nice friendship even though there was a big difference in our age. He could have been my son, but it took us

SHE WAS WIRED DIFFERENTLY

both a while before we knew each other's secrets.

One night on my night off, we went to a small café, on Fisherman's Wharf, for fish and chips. When we finished Matt said, "Hey Cappie, I know of a nice coffee shop down here."

So we began walking. It was a beautiful evening. The cool night breeze was blowing toward us off the ocean, and I felt finally a little contentment coming over me. I knew in my soul I was getting a little better.

Inez was not the only thing on my mind anymore. Long periods of time now passed without me remembering that she was gone or wondering where she was, or what she was doing. Sometimes, though, I would still catch a glimpse of someone who looked like her and the pain in my heart would come back again. It was like something inside me was crying. The longing for those days of companionship would come rushing back sometimes; it would so devastate me for the moment, they almost had driven me into giving up and walking into the arms of that black and gloomy ocean. But this night, with the lights flickering on the water, and this new friend walking beside me, I felt I was on my way out of the despair and felt I could make it without Inez finally.

I still wondered though if she was thinking of me. I looked up at the full moon. Was she looking at it too and wondering about me? I had to shake this sorrow and loneliness again and again. I tried to pay attention to Matt's chattering about the place we were headed for.

We reached a little waterfront coffee shop. It was a small, intimate place, and smelled wonderful—like cinnamon and coffee. Little round tables were crowded together, a lit candle on each with two to four chairs placed around them. We took one near the back where a piano was playing. About half of the tables were taken. As soon as we sat down, a very tall, young, good-looking man was at our table. He offered us a large piece of hand-printed cardboard listing all the different coffee drinks available and descriptions of coffee cakes and pies. Each one included a special name, and the prices ranged from five cents up to ten cents. "Can I bring you fellas some water?" he asked as he set down the menu.

HATTIE

Matt looked up at him and smiled. "Hi Virgil, how's it going? Busy?" The waiter looked at Matt and said, "Oh sorry I didn't recognize you Matt. How are you?" "Fine" Matt replied, not taking his eyes off the guy, and seemed to be waiting for something else to be said. I thought Matt seemed disappointed as Virgil walked away with a gait I was familiar with from my days in the Chicago group. I said to myself, that guy is gay. I knew the walk, and looking back at Matt, it clicked for me! And so is Matt, I said to myself, wondering why it had taken me so long to figure it out. I almost laughed at myself, but knew I had been so preoccupied with my thoughts about Inez that I wasn't aware of what was going on around me.

I thought to myself, I bet he thinks I'm gay too. He looked across the table and said, "I didn't make a mistake about you, did I?" He thinks I'm a gay guy, I thought. I laughed at myself; I really am good in my male role.

It was an awkward moment as I looked back at his open, honest face that now had a sort of puzzled look, "No, Matt," I answered, "no mistake." I leaned across the small table and in a very low voice asked, "Matt, are you gay?"

I was so sure of my instincts by now, I knew I was right. I remembered when I had first met Mel in that cab those many years ago and he had said to me, "We can always detect another cross-gender person." Later he had elaborated on this instinct, one night saying, "We give ourselves away when we let our hair down. We know what we want in a sexual partner. Just like a straight man or woman communicates their desires.

Matt struggled with his answer finally he said, "Yes, I am gay Cappie, you are too aren't you?" Virgil came back for our coffee order. When he left, I said, "Matt, I'm a woman." I felt safe with him. He had been honest with me, so I took the chance of being honest with him. I could tell he was shocked. "My God, I had no idea. You are good!" "Thanks Matt, but I've been living as a man for many years. We are both in the same boat, my friend, just different gender preferences," I laughed.

SHE WAS WIRED DIFFERENTLY

So that evening, down in that little waterfront coffee shop, a new phase of my life began. I didn't know it at the time but Matt and I had formed a bond with one another. The coffee shop became our hangout.

Matt, being younger, attracted a certain gay and lesbian circle of friends. I made friends with some of the older, more mature people who met for coffee. We all felt comfortable and accepted there. It got to be a well-known place for the homosexual community to hang out in the San Francisco area.

With the friends I made in the waterfront coffee shop, I spent long hours in discussion. The companionship answered an inner need of mine. I wrote a little poetry, and listened to some eloquent people express their opinions on society and its misdirected, judgmental opinions about "us."

I came to understand that most families went through a grieving process when they discover their loved ones were so different. It can be like a death. Mothers and fathers think they had somehow caused it. But we couldn't understand their rejection and sadness because to us it felt right. We never knew any other way of being and if our parents shunned us and were embarrassed by us, what did that mean? "If my dear mother or father feels this bad about me, maybe I should just disappear."

I was learning how lucky I had been to have a loving father and mother who loved me the way I was. I decided that God had maybe wired people differently for a reason. What I learned for myself was God seemed to have a purpose for everything. I had met the love of my life and lived happily with her for many years. I felt I had been lucky; many people never had this experience.

I was getting letters from Fred about how the paper and his family were doing. Times were very hard for them. Not only did they have to deal with the depression but also the black blizzard. I heard from Fred in 1938 that their daughter Maye was going to graduate from Rhame High School and that they were getting tired of running the paper. Fred and Nellie were still just barely making ends meet, but things were getting better in the country.

HATTIE

Fred said they were thinking of selling the paper to Mrs. Dahl. I knew her, and knew that she was interested in buying the paper for her son Donald, who had worked for me, years before. In fact, I had taught him the printing business while he was still in high school, but I knew the paper was going to be very hard for him to run without knowing the book-printing business.

I was still not ready to come back so I told Fred to go ahead and sell to them and either rent the house out or board it up. I knew the paper was not making much money for a family to live on. We were lucky anyone wanted to buy it. So it was sold. And we decided to just shut down the book business. I continued on in San Francisco for another year, but my heart I discovered was still in Rhame. Somehow I had come to terms with Inez being gone and had found a spiritual way of thinking about life. I was ready when I got another letter, this time from Mrs. Dahl. She had written me to see if I was interested in coming back and taking back the paper. It was running in the red every month. She just wanted to get out from under the financial burden, and instead of closing the door she thought perhaps I would want it back.

I knew business was picking up around the country. It was 1940; the deep depression was coming to an end. Things seemed to be looking up and so I decided to go back and take over "my" paper again. I was sixty-four and had lived in San Francisco for almost eight years.

When I got back to Rhame, Donald almost ran out of the newspaper office. He said, "You can have it back Cappie, I'm through." So I was back in business again. My health was pretty good. I wasn't depressed anymore. I had found a good partner in God. I asked Donald to stay and help me, and he did.

I opened up my house and cleaned it up. I sat down in my favorite chair and stared at Inez's chair and broke down and cried for the last time as I remembered again the years we had spent in these rooms. She had been gone for eight years.

I could still see her sitting with her legs tucked up under her, reading a book, looking over at me with a sweet little smile and saying

SHE WAS WIRED DIFFERENTLY

something like, "Cappie, do you want a cup of tea?"

When Japan bombed Pearl Harbor on December 7th, 1941, I was putting the paper in the black financially. I started up the book business again.

It was hard on me. I was older now. People would say to me, "How you doing Cappie?" A few old-timers added, "Hear from Inez?" I would always answer with a lie, "Oh she is fine," and nothing else. They soon quit asking about her.

As soon as we declared war on Japan, everything became busy. The whole country seemed to come alive and awake. We were all caught up in defeating first Japan and then the Germans, but it ended up the other way around. All the young in town seemed to leave at once. All the old folks became helpful in any way they could for the war effort. The farmers hired anyone who wanted to work and sold every crop before it was even planted. There was so much news that everyone wanted to read everything. I finally had to hire two young journalists from the high school to help me out. It was a crazy and exciting time, and it was the only thing that kept me going, because I too was caught up in winning the war.

I still thought of Inez, but it was better now. I just hoped she had found a life that made her happy. I wished I knew, but I came to terms with this knowing that God has a plan for us all.

When Roosevelt died, no one was surprised. I liked Truman, even though he was another Democrat. When he ordered the bomb dropped, which really ended the damn war, I was proud of him. It was finally over. Everyone shouted, "Thank God!"

What year is it, I wondered. All of a sudden I remembered, oh, yeah, it's 1946 and I'm seventy. My gosh, how fast the years have gone by. I looked around and saw I was in a hospital bed. I kind of remembered a train ride, and a loud voice calling "Baker, Baker, Montana," and my old friend, Carl, helping me down the train steps.

Then, next thing I remembered was my friend, Doctor Weber, leaning over me and whispering, "You're safe now, Cappie. You're in a private room. The nurse is a friend. I'll be back." Thank God I thought

HATTIE

he is still alive too. He must be in his eighties, good old pioneer stock. I smiled, thanked him with my eyes, and whispered, "Thanks, Doc."

He had warned me about a stroke being worse than a heart attack but that was fifteen years ago. I followed that advice by getting away. I felt pretty woozy and it seemed I couldn't feel part of my body.

I closed my eyes again, knowing I was safe, and thanked God my secret would die with me. My last thought was, "I wonder if a slim little figure would show up at my gravesite?" The date was December 11, 1946.

Census 1920

Papa's Printing Office Helper and Papa "Melroy Fuller" 1910

I had a birthday last Saturday and one of the ladies in town found it out and invited me over to supper ot 6 o'clock dinner, had a swell feed, the only thing lacking was no candles on the cake the reason for the omission being, I think that there wasn't enough in town Ha Ha. Lizzie remembered me with an ivory cribbage board made I think from a walrus tusk, some of the fossil ivory so much of which they find in Alaska.

I see Cappie and Inez once in a while they are getting along fine and Fred and the family were up here not long ago and we all went up to Rhame, I came home the same evening on the train but the rest stayed over night going home the next day. This is the first time Fred has ever visited them and Nellie wrote me they had a dandy visit.

Now you must not think that because your old daddy don't write very often you are ever forgotten, my oldest and I had almost said my best girl, but some of the others might hear of that and you know how jealous they are, but you and yours are often in my thoughts and I hope sometime not far in the future to see you again in the flesh Lovingly MA X

P.S. That cross in the corner is a kiss and don't mean the same as the ones you and Bubba used to put on your letters.

THE WESTERN CALL
AND THE REEDER TIMES
OFFICIAL PAPER OF ADAMS COUNTY — PUBLISHED EVERY THURSDAY
THE FULLER PRINTING CO., PUBLISHERS

Papa's Note

Cappie's Visit to Salem, S.D.
Pal, Marc, Gene, Chris, Hattie as Cappie (far right), Dan and Melroy

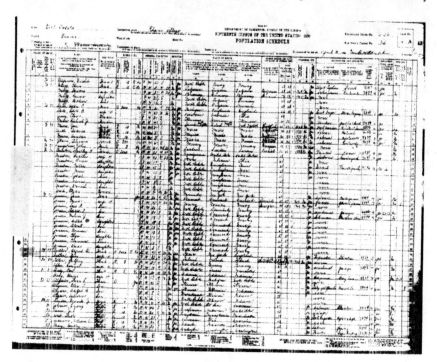

North Dakota, BOWMAN, Roll 1732 Book 2, Page 45

Census 1930

Papa with Author "Lois" Melroy,
Ethyl Tolliver, Elizabeth, Nellie, June, Lois Reimers (age 5)

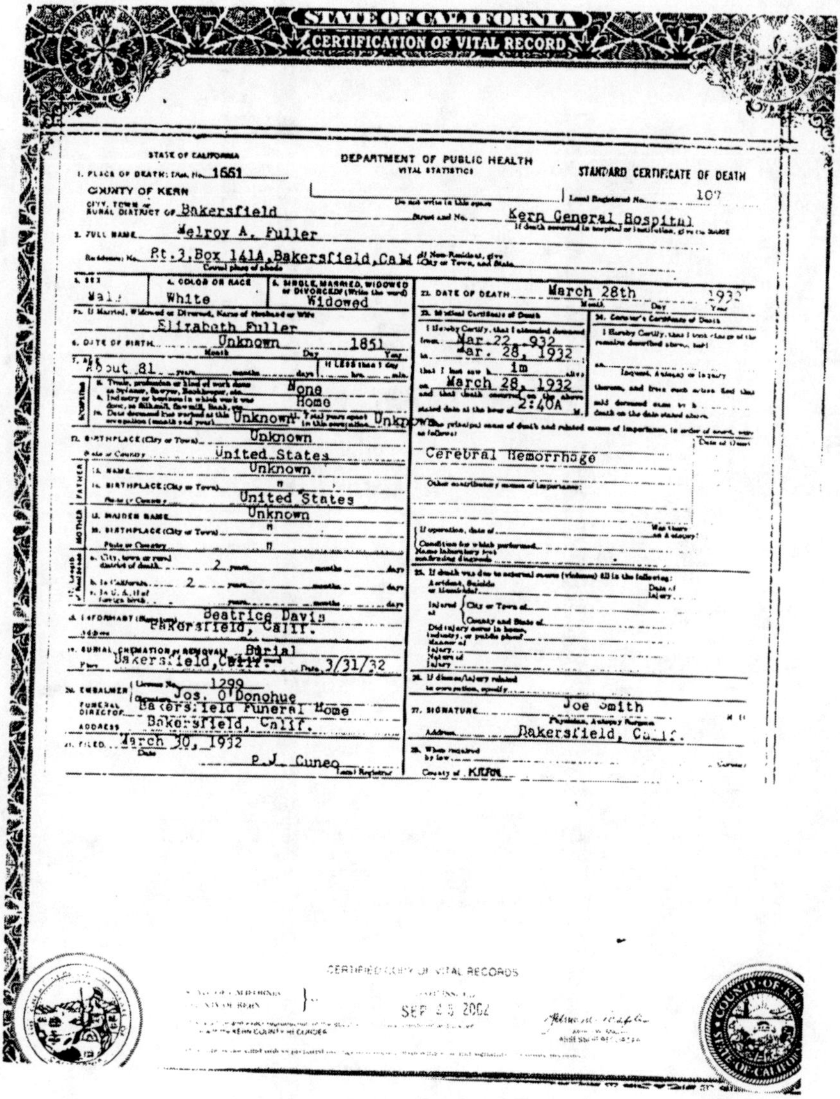

STATE OF CALIFORNIA
CERTIFICATION OF VITAL RECORD

STATE OF CALIFORNIA
DEPARTMENT OF PUBLIC HEALTH
VITAL STATISTICS
STANDARD CERTIFICATE OF DEATH

1. PLACE OF DEATH: Dist. No. 1551
COUNTY OF KERN
CITY, TOWN or RURAL DISTRICT of Bakersfield
Local Registered No. 107
Street and No. Kern General Hospital
2. FULL NAME Melroy A. Fuller
Residence: No. Rt. 3, Box 141A, Bakersfield, Calif.

3. SEX Male
4. COLOR OR RACE White
5. SINGLE, MARRIED, WIDOWED or DIVORCED Widowed
5a. If Married, Widowed or Divorced, Name of Husband or Wife Elizabeth Fuller
6. DATE OF BIRTH Unknown 1851
7. AGE About 81 years
8. Trade, profession or kind of work done None
9. Industry or business in which work was done Home
10. Date deceased last worked at this occupation Unknown
11. Total years spent in this occupation Unknown
12. BIRTHPLACE (City or Town) Unknown
State or Country United States
13. NAME (Father) Unknown
14. BIRTHPLACE (City or Town) "
State or Country United States
15. MAIDEN NAME (Mother) Unknown
16. BIRTHPLACE (City or Town) "
State or Country "
Length of residence: In City, town or rural district of death 2 years; In California 2 years
INFORMANT Beatrice Davis
Address Bakersfield, Calif.
BURIAL, CREMATION, or REMOVAL Burial
Place Bakersfield, Calif. Date 3/31/32
EMBALMER License No. 1299 Jos. O'Donohue
FUNERAL DIRECTOR Bakersfield Funeral Home
ADDRESS Bakersfield, Calif.
FILED March 30, 1932
P. J. Cuneo, Local Registrar

21. DATE OF DEATH March 28th 1932
22. Medical Certificate of Death
I Hereby Certify, that I attended deceased from Mar. 22, 1932 to Mar. 28, 1932, that I last saw h... alive on March 28, 1932 and that death occurred on the date stated above at the hour of 2:40 A.M.
The principal cause of death and related causes of importance, in order of onset, were as follows:
Cerebral Hemorrhage
27. SIGNATURE Joe Smith, M.D.
Address Bakersfield, Calif.
County of KERN

CERTIFIED COPY OF VITAL RECORDS
DATE ISSUED SEP 25 2002

Melroy Fuller Death Certificate

BUFFALO, S.D. NEWSPAPER
(FRED FULLER WORKED THERE UNTIL ABOUT 1916)

The Aaron Groendell families, the la[...] lied by the F. F. F[...] Rhame, left early S[...] ria Bowman. Amid [...] for Medora. Crops [...] man were excellen[...] has exceptionally [...] drive of 20 miles [...] Medora is indeed beautiful. Few drives can compare with it. The road skirts the edge of the Badlands and finally enters them near Medora. Medora is a prettily located village on the little Missouri river in the Badlands, surrounded by trees, hills and fantastic buttes. In the center of the town stands a bronz statue of one of its founders, a well built man with military mien, a low crowned, wide-brimmed hat in his right hand, his left gripping a long barreled rifle, while over his left shoulder hangs a militarly cape. His features are well modeled with a mustache resembling that of Kaiser Wilhelm in his reigning days. A cartridge belt encircles the waist and the figure is equipped with military boots. Such is a brief discription of the Marquis de Mores, French adventurer, dreamer, who left a fine castle and lands in beloved France for the wilds of North Dakota and who met his death by being killed in North Africa. On reverse of the statue are car[...] these words:

In Memory of
Antoine Manca de Vallambrosa
Marquis de Mores
Lieutenant French Cavalry
Born in Paris 1868
Killed in North Africa 1896
and of his wife
Medora
who founded this town in 1883

Taking the west road out of town one passes a tall brick smoke stack, all that remains of the famous packing plant built by the Marquis the rest of the plant having been destroyed by fire. Crossing the river and mounting higher ground one finally gains the plateau where stands he old home of tthis early pioneer, he Chateau de Mores. An American flag flies from the flagstaff and under it the tri-color of France for the Chateau is still wned by the two sons of the Marquis, who reside in France. The mansion is run by Medora people who serve meals and rent rooms o visitors and who guide one rough for 50¢ each. The building is of frame, very large, servants quarters upstairs, sting of 8 or 10 rooms. On ering from the porch, in front e first steps into the living [...] furnished—as [...] rquis. It would take [...] ce to describe the [...] e, fire place, books, Madame Medora, his room. The next st chamber, then the [...] mber. Her bed is en[...] a large canopy which when put[...] in place, was a protection against mosquitoes etc. Another guest room and then the Marquis' office. The maids room, followed then by the dining room, a spacious kitchen etc. One should visit this place for the historical interest it holds.

The Marquis ran a stage line from Medora to the Black Hills, equipped with thoroughbred horses. It is said that he lost around three million dollars in his ventures. Teddy Roosevelt's place is about 16 miles from the chateau.

"DAD" FULLER DIES IN CALIFORNIA

1932

A telegram was received Monday morning stating that M. A. Fuller, father of the writer, had passed away in Bakersfield, Calif., on Easter Sunday, March 27th. Mr. Fuller spent many years in the newspaper business in North and South Dakota, locating in Chamberlain, S. D., in 1881, and coming to the Slope in 1907, where he published newspapers at Reeder and Hettinger and was affectionately known by a host of friends everywhere as "Dad."

The above is from the Rhame Review. Many in this country will remember "Dad" as he visited here occasionally when his son was engaged in the newspaper business in Buffalo some years ago.

A picnic was given at Marmarth Wednesday afternoon for the two Catechism classes of St M[...]schu[...] and thei two eaches s. Miss Agnes Johnson and Miss Maye Fuller. A delightful afternoon was enjoyed by all present, and the teachers were each presented with a lovely gift from their pupils.

Double Envelopes furnished

"A Good Old Bus"—Model 1908

1939

It's Tough Wheeling Now, But The Road Is Better Ahead

With this issue The Rhame Review concludes its twenty-sixth year. It was not our destiny, (nor Rhame's bad luck) that we conduct the paper all of these years, but is has been our fortune to be at the steering wheel for the past seven years, and our misfortune—through the depression.

The Rhame Review:

For twenty-six years The Rhame Review has held its own, willy-nilly, on the precarious edge of the North Dakota Bad Lands — Good Brakes; but the biting blasts of the blizzards of the past few winters have whistled through its pages and stripped them of their advertising —Streamlined, the scorching heat of the past few summers has yellowed its leaves, peeled off its ink and shriveled its perspective — Strong Chassis; but, although unavoidable circumstances have forced it to bow to the inevitable —Knee Action, through all the vicissitudes of grasshoppers and drouth, depression and flood, it has always been able to hold its head above water - Floating Power; and

thank goodness, it is still running a close second to the catalogs in the "two-holers" —Smooth Operation.

Give 'er a little more gas.

A change occurred in business circles Monday when Sanford H. Olson, formerly of Bowman, took charge of the First State Bank as cashier, F. F. Fuller, who has been acting in that capacity for the past two years, stepping out to take charge of his extensive live stock interests in the South Slim Buttes country. Mr. Olson comes here well recommended, having been associated with his brother, Obert A. Olson, in the real estate business at Bowman for the past six years, coming to Bowman from Rochester, Minn. He is a very prepossessing young man, who is a decided addition to our town, and will no doubt acquit himself with credit in his new position.

Melroy Fuller Newspaper Obituary

Handwritten notes on certificate:

Hattie Fuller
Born 1876
Shiefield, Iowa
Died: Dec 11, 1946
Sex F

Death Certification Alfred D. Fuller (Hattie Fuller)
December 11, 1946 Baker, Montana

Alfred D. Fuller "Hattie" Death Certificate

SLOPE MESSENGER

We Cover Slope County Like a Slicker Covers a Cowboy

NEW ENGLAND, NORTH DAKOTA THURSDAY, JANUARY 2, 1947

Traffic Deaths Hit All-Time High

Stark and Cass counties have the dubious honor of having the most people killed in traffic accidents with 12 each in 1946. Cass county had a population of 52,849 and Stark had 15,414 according to the 1940 census. Fourteen counties in the state didn't have a traffic fatality and in this group is Slope, Billings and Sioux in the southwestern part of the state. Golden Valley, Bowman, Adams and Dunn had one each and Hettinger had two.

Total number of people killed in traffic accidents in North Dakota during the past year, according to figures of the Fargo Forum, was 148 and the record was not complete for the year when these figures were released. This was the highest number killed in any one year on record, the next highest being 126 in 1934. Fires claimed the lives of 174 North Dakotans during the past year.

OES-Masons Have Joint Installation

Installation of officers of both Masons and Eastern Stars was held Saturday evening, December 21, at the Masonic Hall. Mrs. Keating presided at the piano during the ceremony and she and Frieda Striebel sang several solos. Mrs. Shirley Richey acted as Installing Marshal. Those installed were Martha Henrickson, Worthy Matron; Shirley Richey, Worthy Patron; Rilla Duckhorn Associate Matron; W. E. Striebel, Associate Patron; Winifred Stuart, Secretary; Laura Swendseid, Treasurer; Verna Spire, Conductress; Dorothy Hedges, Associate Conductress; Bertha Striebel, Chaplain; Frieda Striebel, Marshal; Evelyn Keating, Organist; Marie Bertz, Ruth; Stella Gilmore, Esther; Laura Childers, Martha; Theo. Rushford, Electa; Irene Armstrong, Warder; Fred Armstrong, Sentinel.

The Masons installed the following officers with Shirley Richey acting as installing officer and George Rankin as Installing Marshal: O. G. Hedges, W. M.; Swan Swanson, Sr. Ward; R. C. Rushford, Jr. Ward; A. K. Dahl, Treas.; W. E. Striebel, Se.; George

Give Full Name In Letters To VA

Names such as "Johnson" and "Anderson" top the list of 800,000 National Service Insurance policy holders at the VA Branch No. 8 Office, Fort Snelling, said Wilson Black, contact representative at Dickinson.

A survey of records received at Fort Snelling in the decentralization of National Service Life Insurance shows why this territory is noted as predominantly Scandinavian. In first place are the Johnsons winning with some 9,880. The Andersons "place" with more than 5,600 and the Petersons "show" for a third place tie with the Smiths at 3,300 each.

Running a close fourth are the Olsons tallying 3,000. To balance the field, Jones and Williams also ran, placing fifth and sixth with 1,800 and 1,460 respectively. In seventh place are the Swansons with 920.

Black urged veterans to identify themselves when writing about insurance matters, giving full name, permanent address and insurance certificate number, known as his "N" or "V" number.

County Agent's Weekly Report

There is an increasing demand on the part of farmers for new varieties of grains that will produce high yields every year. To satisfy this demand State and Federal Experiment Stations are spending much time in developing new varieties. Each State Experiment Station works on varieties that will be suited to that particular State. It is true that many varieties that are developed in neighboring states will also do well in North Dakota but it is doubtful if they will be as good for North Dakota as varieties developed by the North Dakota Experiment Station. Another fact that must not be forgotten is that there are many plant diseases as well as troublesome weeds in other states that do not yet occur in North Dakota. Within the past year the Corn Borer is reported to have moved into North Dakota. How it got into North Dakota is not known but a safe bet would be that either feed or seed stocks were to blame. After considering these two points I would hesitate to recommend that seed stocks be bought out of the state. In fact,

A. D. Fuller Dies After Long Illness

On Wednesday, Dec. 11th, A. D. Fuller, former editor and publisher of The Rhame Review, died at the hospital in Baker, Mont., following a long illness.

A. D. Fuller came to Rhame in the latter part of 1916 to look over the situation and become acquainted with the business people of Rhame. On Jan. 1, 1917, he took over The Rhame Review buying the building in which it was housed, the plant and fixtures.

He edited the paper for the next fourteen years, taking part in the civic and school affairs of the town, helping to boost business and giving his best generally to build up the town and surrounding communities. He served for several years as a member of the school board and later as clerk.

About 1930, he sold to his brother, F. F. Fuller helping out at the printing office in rush seasons and running the IOOF Hall. When Mrs. Dahl and Donald purchased The Rhame Review, A. D. helped out frequently with the work at the printing office, then bought it back in 1940, running it until his health failed in 1944, when he sold to the present editor.

Before coming to Rhame, "A. D." as he was known to all the community, worked on the Adams County Record in Hettinger, N. D., having come to Hettinger from Chicago.

Constantly failing health has kept A. D. from taking an active part in affairs during recent years. In September, he suffered a stroke, and had been in the hospital up to the time of his death.

Before his death, he called A. K. Dahl to his bedside on Tuesday, Dec. 10th. He knew his time was short and he requested that there should be no funeral services for him, asking for only a short burial service at his grave.

Rev. K. O. Graves read some passages from the Bible and offered a prayer at the Funeral Home before going out to the cemetery.

Pall bearers were Oscar Carlson, Ralph Perrin, Ole Bakken, Everett Ludlum, Art Johnson and Lee Neville.

His only known relatives are F. F. Fuller and family and a sister, Mrs. D. F. Carey of Hollywood, California.

First Kaiser Car At Local Dealers

The Koppinger Auto Mart Co. of New England has received its first Kaiser automobile, six passenger, 4 door, powered with a 100 HP Continental motor. The car has a roomy interior and very wide seats, motor hood for better visibility and many other features. It sells for slightly over $1900.

George Koppinger and Gabe took delivery on the Linton and flew there in Cessna plane. They were surprised to find that Linton had landing field, but they successfully landed in a wheat field.

The Old Timer

29 YEARS AGO
Taken from the Slope County News, Amidon, North Dakota January 3, 1918

Roland Harding is in Berkeley, California, at the present time studying wireless telegraphy in the aviation department. He listed last summer at Jefferson Mo. Otho Harding also enlisted in the aviation department and is now in training in San Antonio, Texas. Ray Harding was one of the Harding brothers to enlist. Ray went to Omaha, Nebraska and on Dec. 28th enlisted in the aviation department.

E. J. Rotering, the village blacksmith of Midway was in for the first of the week with questionnaire. The people of that way are very anxious that blacksmith not be called into government service until they have their spring work done.

County Auditor and Mrs. Dahl are the happy parents, son, who arrived at their home this village on Saturday, at December 29.

Robert Carr of Ranger, friends in Amidon the first of the week.

G. D. Eaton, assistant cashier at the Bank of Midway, had been accepting treatment at H. C. Towle car hospital several days.

Ole Twedt and Lewis, son, two well known visitors from the Midway country, visitors at the county seat.

Ben Booke from the country was here Wednesday bringing in his questionnaire.

"Slope Messenger" New England, North Dakota January 2, 1947
Story about "A.D. FULLER DIES AFTER LONG ILLNESS"

Hattie Fuller "Alfred D. Fuller" Newspaper Obituary

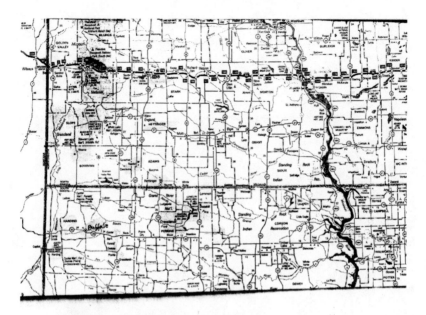

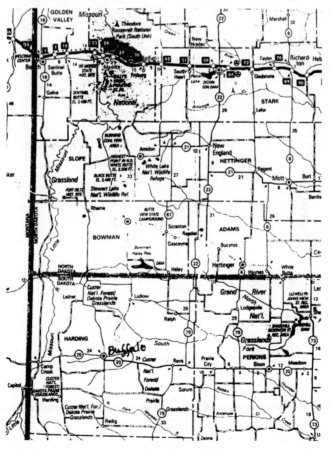